Theodore Cooper

American Railroad Bridges

Theodore Cooper

American Railroad Bridges

ISBN/EAN: 9783337417437

Printed in Europe, USA, Canada, Australia, Japan

Cover: Foto ©berggeist007 / pixelio.de

More available books at **www.hansebooks.com**

AMERICAN RAILROAD BRIDGES.

By THEODORE COOPER.

In all things, but proverbially so in mechanics, the supreme excellence is simplicity.

JAMES WATT.

ENGINEERING NEWS PUBLISHING COMPANY,
TRIBUNE BUILDING,
NEW YORK.

[From Transactions of American Society of Civil Engineers.]

AMERICAN RAILROAD BRIDGES.

BY

THEODORE COOPER, M. Am. Soc. C. E.

The existing and the accepted types of bridges in use to-day on American railroads being the results of a true evolution, no attempt to present them intelligently would be complete without a brief sketch of the past history of bridges in America.

The rapid development of the new world and the enormous number of bridges that has been built within the limits of the nineteenth century, have furnished us with a wide experience, from which we have been able to select the good and reject much that was bad or undesirable.

The pioneer life, not only of the earlier settlers, but of each generation to the present day, has developed to a high degree the energies, ingenuity and self-reliance of the American people. These pioneers were compelled to be men of all trades. Their limited resources and the lack of time or opportunity to seek for past precedents, impelled them to solve each problem anew. They, "thought with vigor and were not fettered with the trammels of science, before they were capable of exerting their mental faculties to advantage," as Sir Joseph Banks wrote to Thomas Paine in 1788.

Having no educated "lines of least resistance," they were better able to solve the many problems before them by new and novel methods.

The bridging of small streams was a part of the pioneers' labor. The crossing of the larger rivers developed specially gifted men, like Timothy Palmer, Theodore Burr, Lewis Wernwag, and others less well known, who built timber bridges that are looked upon as wonderful structures, even to the present day. The records of the early bridges of America are very incomplete, but enough remains to show what admirable work these early bridge builders could do.

1. WOODEN BRIDGES.

The earliest bridges, where single timbers were not sufficient to stretch from bank to bank, were short spans supported on piles, or, where these could not be used, on timber cribs filled with stone. Where the conditions would not allow of structures of this character, arch spans were usually adopted.

In 1660, "The Great Bridge," as it was then called, was built across Charles River, between Old Cambridge and Brighton. It was a pile bridge.

In 1761 Samuel Sewall planned and built a bridge over York River, Maine, 270 feet long, supported on thirteen piers. Rebuilt in 1793.

In 1786 Mr. Sewall built a bridge over the Charles River, at Boston, 1 503 feet long, supported on seventy-five piers. A year or so later bridges on the same plan were built at Malden and Beverly, Mass.

In 1792 Colonel William P. Riddle built the Amoskeag Bridge at Manchester, N. H. It was 556 feet long, and supported on five piers and two abutments. It was commenced on the 3d of August, "at which time the timber was growing, and the rocks dispersed in the river," and completed on September 29th.

Between 1785-92 Colonel Enoch Hale built over the Connecticut River, at Bellows Falls, a bridge 368 feet long, in two spans, taking advantage of a rock in the middle of the river for his center pier. The West Boston Bridge over the Charles River, 3 583 feet long, and supported on one hundred and eighty pile bents, was finished in 1793.

Near the end of the eighteenth century a bridge was built over Cayuga Lake, N. Y. It was a pile bridge in 25 foot spans, one mile in length.

In 1795 a bridge was built over the Mohawk River, 960 feet long, supported on thirteen piers.

In 1792 Timothy Palmer built the Essex-Merrimack Bridge over the Merrimack River at Deer Island, about 3 miles above Newburyport, Mass. It consists of two bridges resting on Deer Island in the midst of the river (Plate I). "An arch of 160 feet span and 40 feet above the level of high-water connects this island with the mainland on one side; the channel on the other side is wider, but the center arch is but 113 feet." That part of the bridge on the Newbury side, the 160-foot span, was removed in 1810 and replaced by a chain suspension bridge. The part on the Salisbury side remained until 1883. The chain bridge was built by John Templeman, of the District of Columbia. "It was the first chain bridge built in New England." Its span is 244 feet between bearings on towers; the towers are timber frames covered with boards and shingles. February 6th, 1827, one of the chains broke under a heavily loaded wagon drawn by four oxen and one horse. The bridge was again rebuilt, and is still in use. The bridge, as it now exists, consists of two independent roadways, each 15 feet 6 inches wide. Each roadway is supported on two sets of chain cables, each set being composed of three chains. These chains seem to have been repaired in many places with different sized links. The chains generally are formed of links about 2 feet long, made of 1-inch square iron. For about 6 feet over the bearings on the towers, each chain is spliced or replaced by three smaller chains with links about 1 foot long, and of about ½-inch square iron. The floor is hung from these chains every 7 feet by suspenders, formed indifferently of bars 1 inch square, straps 2 x ¼ inch, or pieces of chains (Plates II and III).

In 1793 Timothy Palmer built another bridge over the same river at Andover. Rebuilt in 1803.

In 1794 he built the Piscatauqua Bridge, 7 miles above Portsmouth, N. H. It was 2 362 feet long. The greater part consisted of pile work.

"But that part which engages the attention of travelers, is an arc nearly in the center of the river, uniting two islands over water 46 feet deep. This stupendous arc of 244 feet chord is allowed to be a masterly piece of architecture, planned and built by the ingenious Timothy Palmer, of Newburyport, Mass."

"The chord of this arch is 244 feet 6 inches. The versine of the arch was 27 feet and 4 inches, and depth of the frame-work of the arch

18 feet and 3 inches. There are three concentric ribs, the middle one carrying the floor of the bridge; they were selected from crooked timbers, so that the fiber might run nearly in the direction of the curves, and are connected together by pieces of hard and incompressible wood, with wedges driven between, the ribs being mortised to receive them; thus the ribs are kept at a regular and parallel distance from each other. Each rib is formed of two pieces laid side by side about 15 feet in length; they are all disposed in such a manner as to break joints, the end of one timber coming in the middle of the length of the other which is near it; their ends all abut with a square joint against each other and are neither scarfed nor mortised, the two pieces of timber being held together by transverse dovetail keys and joints; all the timbers are admirably jointed and freely exposed to the action of the air; any piece may also be removed in case of its requiring separation without injury to the rest of the structure."

The bridge was 38 feet wide and had three arched trusses. Another description states that the second rib carrying the floor of the bridge was of a larger radius than the lower to facilitate the traveling; the upper rib served for a railing.

In 1794 the bridge at Haverhill was built by Timothy Palmer, "contemporaneous with the first stage coach and the first newspaper." It consisted of three arches of 180 feet; each is supported on three handsome piers 40 feet square; it had as many defensive piers or sterlings extending 50 feet above and a draw 30 feet over the channel.

In 1796 he built a bridge over the Potomac River, at Georgetown.

In 1796 Rufus Graves built a bridge over the Connecticut River, at Hanover, N. H., consisting of a single arch of 236 feet. He patterned his bridge after the Piscatauqua Bridge built by Palmer. The roadway followed the line of the arch and was some twenty feet higher at the center than at the abutments. The bridge was formed of the largest selected white pine, hewed to 18 inches square, some of them 60 feet long. This bridge fell in 1804, without warning, and by its own weight; its destruction being hastened by the undermining of one of the abutments, through deficient waterway. The builder of this bridge studied divinity, graduated at Dartmouth College, became a merchant, then a bridge builder, then an officer of the United States Army, and finally a physician.

In 1795 a bridge was erected at Holt's Rock, between Newbury and

PLATE I.   ESSEX-MERRIMACK BRIDGE.   1792.   Vol. XXI, p. 6.

PLATE II.   ESSEX-MERRIMACK BRIDGE. 1810.   Vol. XXI, p. 6.

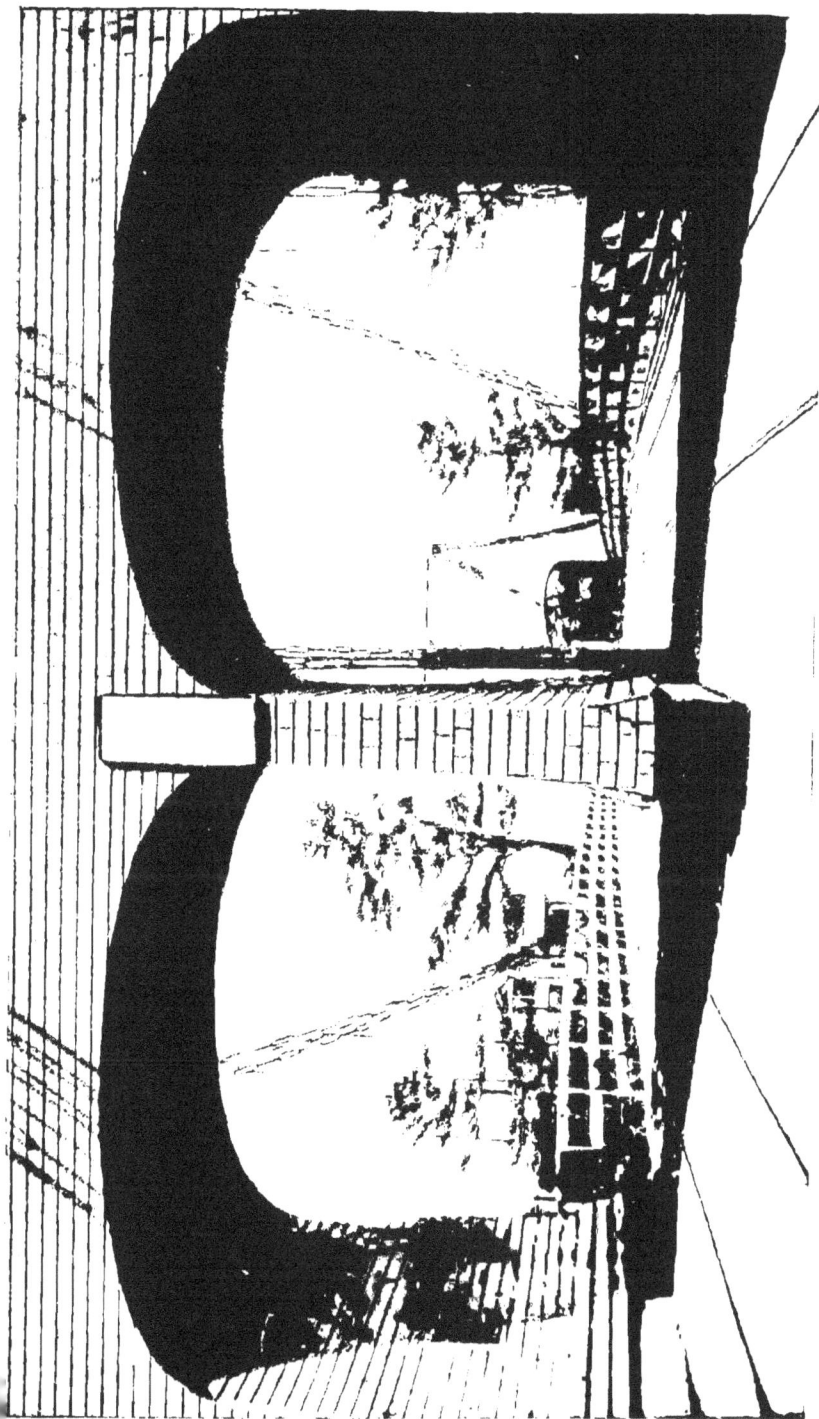

PLATE III.

ESSEX-MERRIMACK BRIDGE. 1810.

Haverhill. It was 1 000 feet in length and consisted of four arches and one draw-span. It was carried away by the ice in 1818.

In 1796 a bridge was built between Harlem and Morrisania, over the Harlem River.

About the end of the last century a bridge was built at Windsor, Vt., with two spans of 144 feet, and one at Romley, Mass., consisting of 8 arches of a total length of 870 feet; one at Howland's Ferry, R. I., 900 feet long, with a sliding draw, supported on 42 pile bents, and the Weybasset bridge at Providence, R. I., "160 feet long, supported by two wooden trestles and two stone pillars."

There were also two bridges over the Lehigh River, one at Bethlehem and one at Easton.

These are the records as I have been able to gather them, of bridges of any importance built in the eighteenth century.

From 1804 to 1806 "The Permanent Bridge" over the Schuylkill River at Philadelphia was built. It consisted of two arches of 150 feet clear and one of 195 feet clear.

"The plan was furnished by Mr. Timothy Palmer, of Newburyport, Mass., a self-taught architect. He brought with him Mr. Carr as his second and four other workmen from New England. They at once evinced superior intelligence and adroitness in a business which was found to be a peculiar art, acquired by habits not promptly gained by even good workmen in other branches of framing in wood." "The frame is a masterly piece of workmanship, combining in its principles that of king post and braces or trusses, with those of a stone arch."

There were three truss and arch frames. The timber was of the best white pine. Width of the bridge, 42 feet. The bridge was covered and closed in from the weather. Mr. Palmer stated that from his experience, wooden bridges uncovered would become unsafe in ten to twelve years.

"I am an advocate for weather boarding and roofing, although there are some who say it argues much against my own interests; notwithstanding I am determined to give my opinion as appears to be right. It is sincerely my opinion that the Schuylkill Bridge will last thirty and perhaps forty years if well covered. You will excuse me in saying, that I think it would be sporting with property, to suffer this beautiful piece of architecture (as you are sometimes pleased to call it), which has been built at so great expense and danger, to fall into ruin in ten or twelve years."

Plate IV is copied from an old engraving. The superstructure was entirely renewed on a more modern plan of a wooden arch bridge in 1850, and widened for also carrying a railroad track. The foundations for this bridge were very difficult of execution, owing to the lack of experience at that day in work of this kind. Depth of water at east pier 24 feet, and at west pier 41 feet, rock bottom with slight covering. The foundations were laid in coffer-dams. Mr. William Weston, an eminent English engineer, who was in Philadelphia at that time, furnished the plan for the coffer-dam; but "declared that he should hesitate to risk his professional reputation on the event." Afterwards, in a congratulatory letter to the President, he says:

"I most sincerely rejoice at the final success that has crowned your persevering efforts in the erection of the west pier; it will afford you matter of well founded triumph when I tell you that you have accomplished an undertaking unrivaled by anything of the kind that Europe can boast of." "I have never, in the course of my experience or reading, heard of a pier founded in such a depth of water on an irregular rock affording little or no support to the piles."

This pier, however, was not carried to the rock, for the weakness and leakage of the coffer-dam became so threatening, the masonry was started three or four feet above the rock.

In 1805 Timothy Palmer completed the bridge over the Delaware River at Easton, Pa. It consists of three spans, 163 feet clear. This bridge, after eighty-four years' service, is still in use and about five-sixths of the original timber is in good condition. It has always been covered in from the weather. (See Plate V.)

In 1804, Theodore Burr built the Waterford Bridge over the Hudson River, consisting of four arch spans, one 154 feet, one 161 feet, one 176 feet, and the other 180 feet, clear spans. The timber is hewn yellow pine. This bridge was not covered until 1814, but is still in use, and in reasonably good condition. (See Plate VI.) The footings of some of the arches required renewal a couple of years ago. The drawings of the Easton and Waterford bridges are made from sketches taken from the existing structures. Burr also built another bridge over the Hudson at Fort Miller.

A bridge over the Connecticut River at Springfield was built about this same date by Mr. Walcott.

From 1804-06, Theodore Burr built the bridge over the Delaware River at Trenton (Plate VII) consisting of five arch spans, two of 203 feet, one of 198 feet, one of 186 feet, and one of 161 feet in the clear. Each span had five arched ribs, formed of white pine plank, from thirty-five to fifty feet in length and 4 inches thick, repeated one over the other, breaking joints, until they formed a depth of 32 inches. The lower chord was composed of two sticks 6½ x 13½ inches. The roadway was suspended from the arch ribs by vertical chains. The arch was counter-braced by diagonal braces, formed of two sticks 6 x 10 inches spiked to the lower chord and secured to the arch above by iron straps. Outside of the exterior ribs wing arches 50 feet long, splayed out so as to widen the bearing on the piers and abutments. The bridge had no other wind bracing. This bridge is erroneously credited to Lewis Wernwag. The arch footings required renewal in 1832, owing to decay of the timber. In 1848 the bridge was remodeled by removing the wing arches and adding a new and stronger arch rib on the south side, and at the same time strengthening the adjacent old arch rib, by increasing its depth to 4 feet, to carry a railroad track. In 1869 it was again strengthened. It was finally removed in 1875, and replaced by an iron structure.

This bridge may, therefore, be considered as a connecting link between the early wooden long-span bridges for highway traffic and those for railroad purposes.

In 1808, Theodore Burr built the bridge shown on Plate VIII, at Schenectady, N. Y. This was the second bridge built at that place by Burr, the first falling before or soon after completion. The traditions in regard to the first bridge are not very reliable. The contract for the one shown, however, states that it is to be "exactly on the site of the bridge built by the said Theodore at the same place," and apparently with different spans, as new piers are provided for, by the contract. There may, therefore, be some truth in the tradition, that the first bridge was to consist of two spans of 450 feet; that one span was completed and the other partially, when it fell or was swept away by floods.

The bridge shown is a curiosity in bridge construction. It is a sus-pension bridge of wood. The curved ribs are formed of eight 4 x 14-inch planks spiked or bolted together. The planks are cut to long bevels at the ends for splicing. The ribs are three in number, spaced about 13 feet apart. The timber was white pine. The plan shown is copied from Burr's original drawing attached to the contract.

The bridge was used for twenty years in this condition. It then became necessary to build additional piers under the middle of each span to stop the excessive sagging of the bridge. In this condition it stood till 1873, when it was replaced by an iron bridge.

From 1812–16, Theodore Burr built the bridge at Harrisburgh, Pa., over the Susquehanna River, consisting of twelve spans of about 210 feet. The half of this bridge which is south of the Island still remains in use.

In 1812 Lewis Wernwag built his " Colossus " bridge over the Schuyl-kill at Fairmount, Philadelphia. (See Plate IX.) It had a clear span of 340 feet 3½ inches. The eastern abutment was founded on the rock, and the western one on piles driven to the rock. This bridge was destroyed l y fire, September 1st, 1838.

In 1814, he built the New Hope Bridge, Plate IX, over the Delaware River, consisting of six arch spans of 175 feet ; versed sine of arch, 13 feet. Each span consisted of three arch ribs, formed of four sticks 6 x 15 inches, dressed to the curve and confined by iron clamps and bolts.

Wernwag's practice was to saw all his timbers through the heart, to detect unsound wood and permit good seasoning of the timbers. He used no timbers of a greater thickness than 6 inches, and separated all the sticks of his arches by cast washers, to allow free circulation of the air.

In 1810 he built a wooden cantilever bridge over the Nashammony River, Pa. He called it his "Economy" bridge, and claimed that it could be used to advantage up to 150 feet spans.

From this date to 1836 he built many bridges. In 1830 he built his first railroad bridge at Manoguay, on the Baltimore and Ohio Railroad.

His last bridge was the railroad bridge over the Canal and Potomac River, at Harper's Ferry, in 1836.

The previously mentioned bridges with spans of any magnitude appear to have been arch bridges.

The first marked step toward bridges of the modern truss form was the wooden lattice bridge, patented by Ithiel Town, January, 1820. Plate X.

The horizontal members or chords of this bridge were composed of two or more parallel sticks, spaced so that the diagonal web mem-b rs or lattice work would pass between them. The timbers were usu-ally 2 to 4 inches in thickness, and 10 to 12 inches wide. The web mem-bers were placed closer together for the longer spans. The chords were also made in two sets for the longer spans, one placed above the other.

The web members at every crossing and at their junction with the chord pieces were fastened with wooden treenails.

The timbers being of one size and of reasonable dimensions, were readily obtainable. The absence of all bolts, straps or rods of iron, and the simplicity of the mechanical operations required to connect together the parts of this bridge, made it a cheap and popular structure.

A great number of bridges of this class, up to spans of 220 feet, were built throughout the United States, both for highway and railroad purposes. There are many of them still in existence.

These bridges were made of uniform section throughout. In order to get sufficient support at the ends, they were usually extended over the abutment a distance about equal to their depth, and were made continuous over all the piers.

From the thinness of the web system the trusses were apt to warp, and as they aged they became very flexible, owing to the want of rigidity of the treenail connections.

Mr. Town claimed, at least in his later pamphlets (1831), that these trusses could be made of wood, wrought or cast-iron.

The next step toward simplicity and concentration of parts appears to have been the truss known as the Long truss, patented by Brevet-Lieutenant-Colonel Long of the United States Engineers, in March, 1830, and November, 1839.

This also was a form of truss in which iron did not enter as a necessary part, the connections being made by framing the parts together, or by use of wooden keys or treenails. Many bridges were built of this form, but it never became widely popular.

Both the Lattice and the Long bridges were combined with the arch in many cases, especially for the longer spans.

No further important advance in the styles of the trusses was made until 1840, when William Howe patented the truss which was the basis for the present Howe truss bridge. Plate XI.

This form of truss grew rapidly into favor, from its simplicity of construction, perfection of detail and satisfactory action under service. For some years it has been the standard form of wooden bridge in use upon our railroads.

The chords and braces are made of timber, and the vertical web members (tension) of round iron, with screw ends, usually upset to give a stronger section through the screw threads than in the body of the rods. The chords are made of uniform section throughout, and both the top and bottom chords of the same number of timbers, and timbers of the same width and spacing. The sticks of the lower chord (for through bridges) are made deeper than those of the top chord, on account of the loss of section from splicing, the lower strains used in tension and to resist the bending strains due to the cross floor timbers, which rest directly on the lower chord. The chord sticks are spaced two or more inches apart, to allow the passage of the vertical rods without cutting away any of the timber, to give room for the packing and splice blocks, and to admit free circulation of air between the timbers. The number and length of the chord timbers are arranged so that no more than one stick need be spliced in any one panel of the truss. The chord sticks may be spliced either with oak or iron splices.

The braces and counterbraces are square-ended, and rest upon the inclined faces of the angle blocks. They are held from displacement by lugs or dowels. The angle blocks were formerly made of oak, but now are made of cast-iron. They are provided with square faced projections, which are let into the chord timbers, to take up the horizontal shear of the braces by direct fiber compression.

The "gib plates," as the plates on which the nuts of the web rods bear are called, are usually made of thick, flat plates of wrought-iron.

The space between the chord timbers below the angle blocks contains cast tube blocks, made a little short to allow for the shrinkage of the timber.

The vertical shear of the braces is transferred through the angle blocks and these tube blocks to the gib plates and web rods, thus avoiding any crushing action across the fiber of the wooden chord pieces.

The timber in a properly designed Howe truss is only strained longitudinally with the fiber.

The very wide experience in this class of bridges has enabled the American bridge builders to so proportion them in all their details that the best results can be obtained from the material at a moderate cost. But for the two objections, their liability to destruction from fire or from decay, no better railroad bridge up to 150-foot spans could be desired. For most of our country the cost of good timber makes iron bridges the cheaper, at first cost, for spans over 150 feet.

Very few Howe truss bridges are now being built upon the older and financially better roads, all wooden bridges being replaced with iron structures.

In April, 1844, Thomas W. and Caleb Pratt patented the truss known as the Pratt truss. It differed from the Howe in making the diagonal members of the web system of iron (tension) and the vertical members (compression) of wood, the reverse of the Howe principle. While many bridges were built upon this form, it never succeeded in attaining an equal popularity with the Howe; as it required a greater quantity of the more expensive material, wrought-iron, and was not so well suited to the joint use of the two materials, wood and iron.

It became, however, the favored form, afterwards adopted, for iron bridges, and is therefore one of the steps in the development of American bridges.'

In ad lition to the general forms of structures heretofore mentioned, the variations of these broad types and other special forms were numerous. Each individual builder had his own favorite at some time. It would be impossible and useless to take up those numerous individual cases, though they undoubtedly served a good purpose in the general evolution.

The development of wooden bridges was entirely empirical. They were generally of uniform section, and their proportions determined either from the result of previous failures or from the study of models. The advancement was, however, toward simplicity of form and construction, and greater perfection of the details of the connection of the parts.

2. IRON BRIDGES.

As early as 1786 the need of a material for long span bridges, more suitable than wood, engaged the attention of thinking men.

Thomas Paine, in 1803, wrote as follows, giving an account of his advocacy and efforts toward the construction of iron bridges:

"As America abounds in rivers, * * * I turned my attention, after the Revolution was over, to find a method of constructing an arch that might, without rendering the height inconvenient or ascent difficult, extend at once from shore to shore over rivers of three, four or five hundred feet and probably more. The principle I took to begin with and work upon, was that the small segment of a large circle was preferable to the great segment of a small circle. The architects I conversed with in England denied the principle; but i· was generally supported by mathematicians, and experiment has now established the fact."

It should be borne in mind that the only iron bridge then in existence in England was the cast-iron arch over the Severn near Coalbrook Dale; which arch was nearly a semicircle, having a span of 100 feet and a versine or rise of 45 feet.

Between 1786–87 Paine made three models, "one in wood, one in cast-iron and one in wrought-iron connected with blocks of wood representing cast-iron blocks." This is Paine's wording; but as the Wearmouth Bridge was copied after Paine's model or bridge afterwards built from this model, it would be clearer to say, the last model consisted of wooden blocks representing cast-iron voussoirs, spliced together by wrought-iron bands fitting into recesses in the blocks and secured by screws.

This last model he took to Paris in 1787 and presented it as a model for an arch bridge of 400 feet span to the Academy of Sciences for its opinion. Its committee, consisting of M. Le Roy, Abbe Bosson and M. Borde, the last two celebrated mathematicians, reported that an arch upon the principle and construction of the model might be extended to the span proposed—400 feet. Paine then had a couple of arch ribs of 90 feet span and 5 feet rise cast at the foundry of Messrs. Walker, at Rotheram, Yorkshire, and tested with double their own weight. On the success of this experiment Paine had a complete bridge of five arch ribs of 110 feet span and 5 feet rise made by the same founders and sent it to London, " as a specimen for establishing a manufactory of bridges to be sent to any part of the world." Soon after this his interest in bridges was obliterated by his interest in the French Revolution. His bridge was sold to satisfy his creditors.

In the memoir from which the above information is taken, dated 1803, he proposes that the Congress of the United States erect an experimental arch of 400 feet span "to remain exposed to public view, that the method of constructing such arches may be generally known." He offers to furnish the proportions of the several parts and to superintend its construction without compensation.

From this time no further consideration seems to have been given to the use of iron for bridges until about 1830, when Long and Town both suggested that their bridges could be made in iron or wood.

In 1833 August Canfield took out the first patent for an iron truss bridge.

The first iron truss bridge built is believed to be the one erected over the Erie Canal at Frankford, N. Y., in 1840, by Mr. Earl Trumbull.

" The truss was a combination of the truss and suspension principles, and was formed of—first, seven cast-iron sections or panels of about 11 feet in length and 7 in depth, cast solid, each segment consisting of an upper chord, a pair of diagonal braces and half of a hollow cylindrical post at each end (of the segment), except that the end segments had full cylindrical posts at the abutments. These semi-cylinders being bolted or clamped together in series, formed full cylindrical posts, which were flanged at the bottom, and through which were passed vertical bolts securing them to wooden transverse floor beams. Second, two wrought-iron suspension rods (1¼ inches diameter) attached to the top end of the end posts, and sagging in a parabolic curve, so as to pass under and support the two centermost floor beams, and under lugs cast at proper elevations upon the posts intermediate between the centermost and the end posts, whereby such intermediates were supported. Cross-sections of chords and diagonal braces of the + formed section."

In the same year (1840) Mr. Squire Whipple built his first iron bridge. It was a bow-string truss bridge in which the compression members were of cast-iron and the tension members of wrought-iron. He took out Letters Patent for this style of bridge April 24th, 1841. A large number of bridges on this plan with spans from 50 to 100 feet have been built, and some few with spans as long as 180 feet.

In 1846-47 James Millholland built a boiler plate tubular girder 55 feet long for the Baltimore and Ohio Railroad at Bolton depot.

In 1846 Frederick Harbach patented an iron Howe truss; the top chord and braces were of cast-iron and the lower chord and vertical rods were of wrought-iron. Each chord and the main braces were hollow cylinders in pairs, the lower chord was of boiler iron riveted together as continuous tubes. The braces and chords were connected by cast-iron saddles shaped to fit the chords.

A bridge of this style, 30 feet clear span, 6 feet depth and 4 feet panels was built in 1846-7 on the North Adams branch of the Boston and Albany Railroad near Pittsfield, Mass.

About 1847-50 Nathaniel Rider built a number of bridges for the New York and Harlem, Erie and other railroads. The Rider bridge was composed of parallel top and bottom chords and multiple systems of vertical posts and diagonal ties. The top chord and posts were of cast-iron; the lower chord and ties of flat bars of wrought-iron. A wedge was inserted at the top of each post to tighten up the systems. The members were bolted together. The failure of one of these Rider bridges on the Erie Railroad in 1850, following the failure in England of the bridge over the River Dee, influenced the officials of that road to the decision that iron bridges were untrustworthy, and to direct that all iron bridges, consisting of several Rider bridges and some which had been built by Squire Whipple, be removed and replaced with wooden structures.

"The first impulse to the general adoption of iron for railroad bridges was given by Mr. Benjamin H. Latrobe, chief engineer of the Baltimore and Ohio Railroad. When the extension of this road from Cumberland to Wheeling was begun he decided to use this material in all the new bridges. Mr. Latrobe had previously much experience in the construction of wooden bridges in which iron was extensively used; he had also designed and used the fish-bellied girder constructed of cast and wrought-iron; he had adopted on the older portion of that road the Bollman plan of bridge for short spans. For the bridges west of Cumberland he adopted the plans submitted by Mr. Albert Fink, his assistant."

In 1852 a span of 124 feet upon the Bollman plan was completed at Harper's Ferry. Bollman's truss is shown on Plate XII.

In 1851-52 three spans of 205 feet each were built over the Monongahela River upon the Fink plan. Plate XIII.

Both of these styles of bridge were of the suspension truss form, being different developments of the trussed beam.

The chords and posts were of cast-iron and the tension members of wrought-iron.

The wrought-iron tension bars had eyes at the lower ends and were connected to the feet of the posts by pins. At the top these bars had screw ends for adjustment. Plates XIV and XV show other forms of the Fink truss.

In 1852-53 Squire Whipple built upon the Rensselaer and Saratoga Railroad, 7 miles north of Troy, N. Y., an iron bridge of 146 feet clear span. (See Plate XVI.)

It was a double intersection Pratt truss with inclined end posts, a form which was known afterwards as the Whipple truss. The top chord and posts were of cast-iron. The lower chord was composed of links of wrought-iron. The web rods were round rods of wrought-iron with eyes at the upper end and screw ends below. They connected to a pin through the top chords and pass through holes in the cast-iron shoe at the foot of the posts. This cast-iron shoe had oblong trunnions at each side to receive the lower chord links. Although this bridge was proportioned for a rolling load of one ton per lineal foot only, it continued to do good service until 1883, when it was taken down, as it was deemed prudent to replace it with a bridge of wrought-iron adapted to the requirements of the then existing traffic (the engines in use upon the road having increased in weight from engines of 35 to 40 tons to engines of 50 to 68 tons).

A few months before this bridge was taken down one of the pair of main web rods 1¼ inches diameter broke with a "suspicious looking fracture;" the remaining rod, however, had sufficient strength to prevent a collapse of the truss.

From 1851–61 many iron bridges from 65 to 110 feet spans were built on the western and mountain divisions of the Pennsylvania Railroad. There were Pratt trusses stiffened with arches, the top chords, posts and arches being of cast-iron. They were built at the company's shops.

In 1857 Mr. F. C. Lowthorp built his first railroad bridge on the Catasauqua and Foglesville Railroad, Pennsylvania. It consisted of 11 spans, total length 1122 feet, supported on iron trestle towers 89 feet high. The chords and posts of this bridge were of cast-iron and the tension members of wrought-iron with adjustable screw ends.

About 1856 John W. Murphy built a Whipple bridge over the Saucon Creek for the North Pennsylvania Railroad, of 152 feet clear span. The road was never completed on this particular line, and the bridge still stands as an isolated specimen of one of the earliest forms of iron railroad bridges of the Whipple type.

In 1858–59 Murphy built for the Lehigh Valley Railroad a Whipple-Murphy bridge of 165 feet span, over the Canal at Phillipsburg, N. J. In this bridge he substituted for the cast trunnions on the post feet of the Whipple bridge pins of wrought-iron, unturned. The lower chord was formed of wrought-iron elongated links similar to the Whipple form. The main web bars were wrought-iron bars with looped eyes at each end. The counter bars were also bars with looped eyes, but the lower eye was elongated and fitted with gib castings and keys for tightening the bars.

This is the first truss bridge, as far as the author has been able to discover, which was pin connected throughout.

In 1869 this bridge was taken down, and put up as the middle span in a long wooden bridge of nine spans, at Towanda, Pa., to reduce the liability of destruction of the whole bridge by a fire.

In 1879 it was again removed, and rebuilt by substituting wrought-iron compression members for the cast-iron, turning the pins, reboring the links, etc., and put up at Shepherd's Creek, on the Southern Central Railroad.

In 1861 Mr. J. H. Linville built a bridge on the Delaware extension of the Pennsylvania Railroad over the Schuylkill River, in which were used for the first time wide forged eye-bars and posts formed of wrought-iron sections.

In 1863 John W. Murphy built a bridge over the Lehigh River at Mauch Chunk for the Lehigh Valley Railroad, in which he used wrought-iron for both the posts and top chord sections. This is probably the first American truss bridge in which the tension and compression members were of wrought-iron. He used, however, cast-iron joint blocks and pedestals.

In 1865 Mr. S. S. Post built the first iron bridge of the style since known as the Post Truss, for the Erie Railroad, at Washingtonville, on the Newburgh branch.

Between 1865-80 a large number of bridges on this style were built. From its peculiar form, the counter system of web members consisted of a single system, which met at the center a double sytem of direct web members, thus rendering the strains ambiguous. Much of its early popularity was due to the apparent stiffness of this form of truss under moving loads—the truss being put under an initial strain by the counter system which extends the whole length of the truss.

In 1859 Mr. Howard Carroll commenced building riveted lattice bridges for the New York Central Railroad and its connections. The bridges built by him were mostly short spans, varying from 40 to 90 feet. They were well proportioned both in strength of parts and in the detail of the riveted connections, for the service then considered as sufficient.

This early work of an excellent character gave the preference to this class of bridges upon this road and other neighboring systems.

Mr. Charles Hilton, a pupil of Mr. Carroll, continued his work and extended the system to longer spans, the longest being the bridge built by him over the Connecticut River at Springfield, Mass., in 1874. It consisted of seven spans of 180 feet.

In 1864 Mr. Felician Slataper designed and built for the Pittsburgh, Fort Wayne & Chicago Railroad a bridge over the Allegheny River at Pittsburgh. It consists of five riveted lattice spans, 178 feet each.

In 1865-66 William Fairbairn & Co., of England, built the iron lattice bridge over the Connecticut River on the Hartford and New Haven Railroad, near Windsor Locks, from designs by Mr. James Laurie (Past Prest. Am. Soc. C. E.), the longest span being 177¼ feet.

Long Span Bridges.

The era of long span truss bridges in America may be considered as dating from the building of the first bridge over the Ohio River at Steubenville, between 1863-64, by Mr. J. H. Linville. The channel span was 320 feet long and 28 feet deep. The top chord and posts were made of cast-iron. It was proportioned for a rolling load of 3000 pounds per foot of track, a notable increase in the load heretofore in use. (See Plate XVII.) This bridge is now being removed.

This was followed by the bridge over the Ohio River at Louisville, built by Mr. Albert Fink in 1868-70, with two main spans of 360 and 390 feet. (See Plate XVIII.)

In 1870 Mr. Linville built the bridges over the same river, at Parkersburg and Bellaire, with channel spans of 340 feet.

And in 1872 he built the Newport and Cincinnati Bridge, with a channel span of 420 feet.

During 1868-74 James B. Eads built the bridge at St. Louis over the Mississippi River, consisting of two arches of 502 and one of 520 feet clear span.

In 1876 the Cincinnati Southern Railroad bridge over the Ohio River was built by the Keystone Bridge Company under specifications prepared by Mr. G. Bouscaren, the design being made by Mr. J. H. Linville. The channel span was 519 feet, the longest truss span ever built up to that time.

The channel spans of the Henderson Bridge, 522 feet, and of the Ohio River Bridge, Kentucky Central Railway, at Cincinnati, 550 feet, are the only independent truss bridges up to the present time with greater spans.

There are now in existence on American railroads over five miles of bridges with spans from 300 to 400 feet, nearly four miles with spans from 400 to 500 feet, and two and a half miles of bridges with spans exceeding 500 feet, estimated as single track bridges, all exclusive of wire suspension bridges.

The first iron cantilever bridge of importance was the Kentucky River Bridge designed and built by C. Shaler Smith in 1876-77. It is

1 125 feet long, and consists of three equal spans 375 feet. The second, also by the same engineer, was the Minnehaha Bridge over the Mississippi, near St. Paul. It was built during the winter of 1879–80, and consists of one center span 324 feet, and two shore spans 270 feet.

The third and fourth were the Niagara and Frazer River Bridges, designed by Mr. C. C. Schneider, the first being completed in 1883 and the second in 1885.

The St. John River cantilever was built in 1885. Since that time many have been built, the one of greatest magnitude being the Pough-keepsie Bridge, built by the Union Bridge Company and already described in our Transactions (Vol. XVIII, June, 1888).

3. COMBINATION BRIDGES.

This kind of bridge is one in which all the tension members of the web and of the lower chord are made of wrought-iron and all or most of the compression members of wood.

In the South and West a great number of these combination bridges have been used. They are cheaper than bridges of all iron parts and are more permanent than wooden bridges, the wooden members being less liable to destruction from fire and decay, where the timber is only used in compact forms and under compressive strains. They are generally designed so that the greater part, at least, of the iron members can be used again, when it becomes desirable to renew them as wholly iron bridges. While more permanent than wooden, they are less so than iron bridges.

The question of the adoption of this class of structure instead of all wood or all iron bridges, must be determined for each locality by the relative economy as determined by their first cost and the financial ability and policy of the railroad.

Their forms are varied according to the preference of different engineers. We illustrate (Plates XIV and XIX) forms designed by Mr. Alfred Fink, and which have been in general use in past years upon many Southern railroads, especially the Louisville and Nashville Railroad.

This road has been rapidly replacing them of late with wrought-iron bridges of modern type. While excellent bridges can be made of this kind, they should only be considered as an advance from wooden bridges towards the future all-iron bridges.

4. DESIGNING AND PROPORTIONING.

The early wooden bridges and some of the early iron bridges were not designed or proportioned upon any very correct theory. They were largely empirical structures, generally of uniform section throughout their length. Practical experience had led to improvements in methods of counterbracing against partial loads, and in providing for better detail, to obviate the crushing of the timber or slipping of the braces and to take up the looseness arising from the shrinkage of the timber.

In the early days of our railroads so able an engineer as Mr. Kirkwood writes:

" The heavy locomotives used in the freight traffic of railroads, and the great speed of express passenger trains, render massive strength in all railroad bridge structures essential. The many bridges which have been strengthened since they were built and the many which have had to be removed and replaced by stronger structures, sufficiently attest the mistaken economy of erecting light and cheap frame bridges. The blind confidence which the best of carpenters are apt to have in the raw material of their trade and in the rude geometry of its usual combinations, has rendered them unsafe guides to engineers, in the planning of bridge trusses involving irregular and shifting strains, with which their general practice cannot make them familiar. Hence it is that, even now, the engineer will always find it not merely a safe but a necessary rule, to make his bridge trusses heavier and stronger than the most intelligent professional carpenter of his acquaintance is likely to consider necessary."

The heavy engines of that day would be very light ones at the present time.

The science of bridge proportioning was yet undeveloped. The best that the engineer could do, was to make the bridges stronger than heretofore, solely on the facts brought out by past experience.

The first attempt to analyze the strains in skeleton bridge trusses appears to be the work of Mr. Squire Whipple, the inventor of the Whipple bridge—the prototype of the present most generally adopted form of truss now in use in America. His book was published at Utica, N. Y., in 1847. It does not seem to have been widely known till a much later date. For Mr. Herman Haupt, who had an extensive experience in building wooden railroad bridges, and whose treatise on bridge construction, published in 1851, was long an authority, states in his preface:

"If any work exists containing an exposition of a theory sufficient to account generally for the various phenomena observed in the mutual action of the parts of trussed combinations of wood or metals, the author has neither seen nor heard of it." " The best works on the subject of construction that have fallen into the hands of the author contain but little that will furnish the means of calculating the strains upon the timbers of a bridge truss or of determining their relative sizes."

While the principal material of construction was wood, its cheapness and the practical reasons for making the principal members of the trusses of uniform section, made the necessity of proper proportioning less apparent. The advent of iron construction, involving the use of a material of a more expensive character, compelled a better consideration of the relation of the principal parts of a truss to the duties to be performed.

Besides the defective knowledge as to the calculation of the strains, there was a like deficiency in regard to the strength of materials in the forms adapted for skeleton structures, and as to the effects of the concentrated engine loads.

The use of a uniform load of one ton per foot of track for designing all parts of a bridge was in general use some years after 1860 and on many bridges up to even a later date. It was not till about 1870 that a heavier uniform load was adopted for the floor system than was used for the main trusses.

The rapid growth of the traffic on our roads and the constantly increasing loads upon our cars and engines were not fully appreciated until very recent times.

Up to about 1874 the designing and the construction of bridges were almost exclusively, in the hands of the several bridge companies. Each of these companies had its own peculiar style of bridge ; consisting either of a special form of truss, as the Bollman, Fink, Linville, Post, etc., or in the use of patented forms, as the Phœnix, Linville & Piper and American Companies' posts, and other detail parts.

Each company also had its own special geographical field, or lines of railroad, giving it the preference.

Even at points where they did meet as competitors, it was rather as advocates for their special trusses or forms of parts, than as competitors upon any very definite specification.

Though many excellent bridges, considering the state of the art, were built under this system, there were also many very inadequate structures made.

The construction of the St. Louis Bridge marked a most important advance in the development of American bridges. The investigations made, during the building of the St. Louis Bridge, into the strength and other properties of the materials of construction, and especially the testing of full-sized members and their detail connections, not only advanced very much our knowledge of these matters, but also gave an impulse to such investigations which has continued to the present. The erection of the St. Louis arches, by building out from the piers, was the first practical solution of the cantilever principle on a large scale. The erection of two balanced cantilevers, each over 250 feet, with ease, safety and economy, made clear to the mind of engineers that the cantilever was the economic method of erecting long spans over deep gorges or rivers, where ordinary methods of scaffolding would be too expensive, or subject to great risks, or where navigation forbids the obstruction of the waterway.

The construction of the Cincinnati Southern Railroad also marked an important step in the improvement of bridges and their construction.

The bridges upon this road, including the large span at Cincinnati over the Ohio River, and the crossing of the Kentucky River gorge, comprised the first work of magnitude that was offered for competition upon specifications drawn by an engineer acting exclusively in the interest of the Railroad Company.

Two points in these specifications of Mr. Bouscaren were especially important ones : 1st. The use of an engine and train diagram for the live load, instead of the prevalent method of a fixed uniform load per lineal foot of bridge or floor. 2d. The requirement that full-sized compression members of the different competitors be submitted to test, to determine their ultimate strength and elastic limit.

About 1874, the bridge engineer, acting in the interest of the railroads, began to exert a beneficial influence. New competitors as bridge contractors were also entering the field.

These influences all worked together to make a great improvement in the general character of the structures built after that date.

The failure of the Ashtabula Bridge in December, 1876, not only alarmed the general public, but also shook the blind confidence of the railroad companies in their existing bridges.

Upon the principal roads a general examination was made into the character and condition of their bridges by competent engineers with very beneficial results. Not only were many defective structures rebuilt or strengthened, but the defective methods heretofore in use, both in the proportioning of parts and in constructive detail, became apparent.

Floor systems designed for a uniform rolling load, regardless of the length of the panels, were found to be inadequate for the concentrated engine loads in use. For the same reason the counterbracing was also deficient. Broken castings emphasized the unreliability of this material for bridge purposes. A careful examination and analysis of the details, upon which the capacity of the structures depended as much as upon the sectional areas of the several members, showed the need of greater attention to such matters.

The Erie specifications drawn by the writer in 1878, embodying the results of his experience in the designing of bridges, their shop construction, the testing of material and the study of existing bridges and their defects, was the first general specification covering the designing, proportioning and detail of construction with that completeness necessary to give the railroad company the full advantage of the competitive method, with a certainty that the resulting structure would in all ways be up to the advanced state of the art.

It was the first paper on bridge construction in which that relic of ignorance, "the factor of safety," was entirely omitted.

It definitely specified the working strains to be allowed on the different parts of the structure according to the service they were to perform.

While aiming to obtain the best results for the railroad company, by clearly specifying for what loads and under what working strains each part should be proportioned, and the perfection of workmanship which must be adhered to, it left to the contractor perfect liberty of selecting his own plan of structure, as to form of truss, relative proportions of depths and panel lengths and methods of detailing.

These specifications were adopted very widely. They have been modified and extended from time to time, by the author and other engineers, as advancing knowledge and experience justified.

The competitive struggle of the bridge contractors has still been allowed free scope in certain directions, but has been modified by the new factor, the bridge engineer acting in the interest of the railroad companies.

The bridge engineer acting for the contractor has for his principal incentive, the selection of a design which, while complying with the specifications in all ways, can be built at the shops for the least money expended in time and material, and which can be erected at the particular location with the least cost and risk.

The bridge engineer acting for the railroad company, while giving all due weight to the aims of the contractor, is seeking to get a structure which will give the best results after its completion, for operation and maintenance, without undue cost.

As the engineer of the contractor may at some other time be the engineer acting for the railroad and the reverse, these different views are constantly reacting on each other, to the great benefit of our bridge structures.

The competition to-day between the different bridge companies has been largely reduced to the question of shop management, or the relative cost of turning out so many tons of bridge work in a certain limited time.

Practical bridge engineers have given little weight to the theoretical examination of the economical truss. They have recognized as fallacious the deductions usually drawn by theorists from the neglect of most important factors in such investigations. True economy does not necessarily mean minimum weight of material. The relative weight of the various classes of iron in a bridge, the cost of manufacturing each form, rolled, riveted or forged, facility of transportation and erection and other factors must be given their due weight ; and even these factors must change from time to time, and do vary even for the same time at different manufactories. While the tendency towards economy of material due to changes in forms and proportions has been fully recognized, the comparison of alternate or competitive plans has led to the proper proportions of true economy with far greater certainty.

The engineer who is constantly employed in preparing competitive designs, can choose almost instinctively the proper proportions to give his design, selecting the panel lengths, depth of truss and skeleton system best suited to give the proper relations between the floor, web and chord systems of the bridge, with due regard to a suitable arrangement for the lateral and sway bracing.

No fixed proportions can be permanently established suitable for all cases ; for in addition to the varying elements before mentioned, the question of head room, waterway and skew often necessitate special changes. There must also be a certain undefined harmony in the relation of the several parts, in order that the disproportion or inefficiency of the connection of parts may not render the design a defective one.

A successful bridge engineer, from the American point of view, must be something more than a mere calculator of strains. That is the most elementary part of the duty, and does not come within the province of designing. After the selection of the skeleton form and relative proportions of panels, depths and widths of span, a very moderate knowledge of mechanical mathematics would enable any one to determine the strains in an American bridge. He must, in addition to his knowledge as to the effects of varying forms and proportions, have a full knowledge of the capacity of his forms and their connections, and also of the practical processes of manufacture and erection. He must know how his design can be made and put together, and whether it is so harmonized in all its parts and connections, that each part may do its full duty under all possible conditions of service.

In addition to knowing all the elements that make up a perfect design, he must have the instinct of designing or the power of adapting his knowledge to any individual case, in order to obtain the best or the desired result.

Then experience, observation and a sharp competition with men of like knowledge and instinct will give him his position as a bridge engineer.

The equivalent uniform loads given in the following tables (1 and 2), represent in a general way the live loads used in designing bridges during the past, and the maximum live load now being adopted by certain of our railroads.

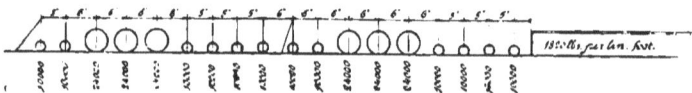

Fig. 1.

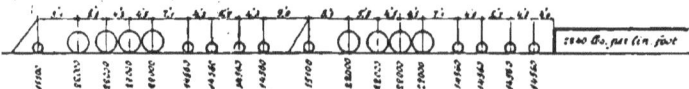

Fig. 2.

TABLE No. 1.

SPANS. FEET.	FROM MOMENTS—LOADS FOR LONGITUDINAL GIRDERS.				
	Uniform Load of One Ton per foot.	Cincinnati Southern Train Load. 1873. Fig. 1.	Erie Train Load. 1878. Fig. 2.	101 Ton Consolidation, followed by 3 000 pounds. 1884. Fig. 3.	Heavy Grade Lehigh, followed by 4 00J pounds. 1889. Fig. 4.
5	2 240	8 000	8 800	12 000	10 000
10	"	4 790	5 290	6 750	9 610
15	"	4 470	5 280	6 670	9 600
20	"	4 320	4 640	6 190	8 560
25	"	3 970	4 400	5 850	7 950
30	"	3 770	4 200	5 470	7 290
35	"	3 540	4 080	5 110	7 040
40	"	3 400	3 850	4 760	6 690
45	"	3 230	3 700	4 650	6 000
50	"	3 140	3 590	4 350	5 870
55	"	3 050	3 450	4 220	5 680
60	"	2 960	3 400	4 110	5 480
70	"	2 920	3 220	3 940	5 110
80	"	2 920	3 260	3 950	4 800
90	"	2 860	3 250	3 970	4 560
100	"	2 800	3 230	3 910	4 520
125	"	2 780	3 150	3 830	4 740
150	"	2 670	3 070	3 740	4 760
175	"	2 590	2 990	3 650	4 670
200	"	2 530	2 910	3 590	4 640
225	"	2 440	2 850	3 520	4 550
250	"	2 390	2 800	3 480	4 470
275	"	2 360	2 750	3 430	4 380
300	"	2 340	2 710	3 390	4 290
350	"	2 240	2 660	3 230	4 250
400	"	2 200	2 610	3 180	4 200
450	"	2 160	2 570	3 160	4 160
500	"	2 120	2 540	3 150	4 120

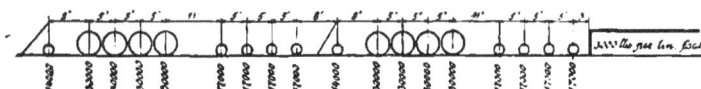

Fig. 3.

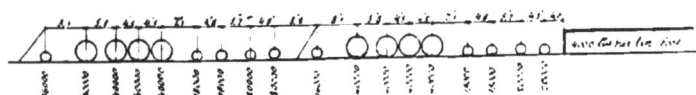

Fig. 4.

TABLE No. 2.

EQUIVALENT LOADS FOR ERIE TRAIN LOAD.

Span—Feet.	From Moments.	From Double Shear.	From Single Shear.
5	8 800	5 280	9 680
10	5 290	4 620	7 200
15	5 280	4 210	6 240
20	4 640	3 830	5 400
25	4 400	3 588	5 040
30	4 200	3 400	4 740
35	4 080	3 270	4 540
40	3 850	3 175	4 300
45	3 700	3 090	4 220
50	3 590	3 030	4 240
55	3 450	3 000	3 930
60	3 400	2 970	3 830
70	3 220	2 960	3 680
80	3 260	2 870	3 620
90	3 250	2 820	3 570
100	3 230	2 750	3 450
125	3 150	2 570	3 280
150	3 070	2 450	3 180
175	2 990	2 400	3 090
200	2 910	2 370	3 000

Table No. 1 gives the approximate equivalent uniform loads for the moments due to several forms of typical live loads upon spans of different lengths. The approximate dates on which these loads were introduced are also given.

These equivalent loads represent in a general way the growth in the required strength of our best bridges, from the earlier days of iron bridge building to the present time.

This table will make clearer than any other method of description the difficulties American bridge builders have had to contend with. In face of the constant progress and bettering of our bridges we have had to meet rapidly increasing demands for greater loads; with too often an increasing reluctance on the part of the authorities to build for anything but the immediate present requirements.

It will be readily seen from this table why our shorter spans are being overtaxed, and why in all our older bridges the floor and counter systems are too weak. The above table represents simply the equivalent loads, which will render the same maximum moments as the corresponding train loads. To provide for the maximum action of any miscellaneous loading, we must also consider the equivalent loads, which represent the maximum shears. For this purpose we need two additional equivalent loads—one to represent the maximum end-shear upon a single span of girders or trusses, which we will call the "single shear;" and another to represent the maximum shear which can occur at the supports of two adjacent spans or panels of girders or trusses, which we will call the "double shear."

The single shear equivalent loads give us the loads on the end members of the trusses and girders. The double shear equivalent loads give us the loads on the supports, whether they be columns or cross-girders.

Table No. 2 gives for one engine the equivalent loads corresponding to the maximum moments, and maximum double and single shears. They differ from one another for each different span of truss or panel. We must therefore use at least three series of uniform loads to represent even approximately one engine load.

The maximum moment of a miscellaneous loading being generally at some other point than the center of the girders, the equivalent load giving the maximum moment will be different for a girder with uniform flanges and for one with flanges of uniform strength.

To represent the action of a train load properly by this method, we should need four different series of equivalent loads, all varying for each span.

The apparent simplicity, therefore, of using equivalent uniform loads for proportioning our structures, is a fallacious one; when applied to partial loads it becomes far more confusing and untrustworthy.

The old method of computing the strains by use of panel loads and equivalent uniform loads (and too often using only the one derived from the moments) has gradually given way to a system much simpler, more accurate and also as easy of application.

The principles and application of the new method were worked out independently but simultaneously by Mr. Robert Escobar, C. E., of the Union Bridge Company, and the author, some years ago (1880).

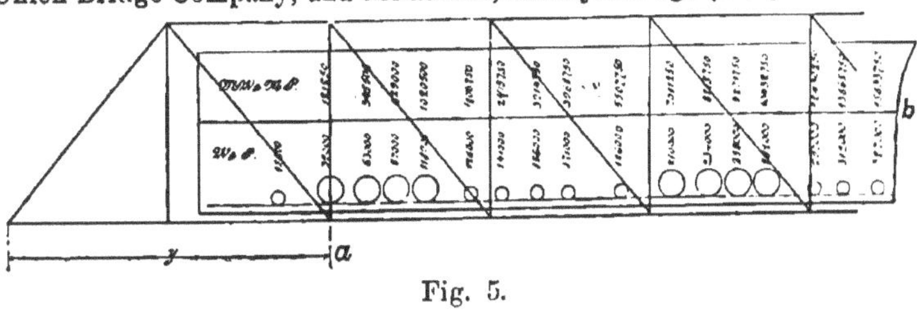

Fig. 5.

Let Fig. 5 represent a skeleton truss of a single system of triangulation and n panels.

The maximum shear at any point a will occur, when one wheel of any miscellaneous series is on the panel point, and when the total weight of train on the bridge, W and n times the loads in front of a (not including the one at a), are the nearest to an equality; or, in other words, calling P the loads in front of a, when $W - n\,P$ is a minimum.

Similarly the maximum moment at any point a will occur when the train is so placed that $W : P :: L : y;$ L being length of span.

Application of the method: Lay down the engine and train load on a card b to the same scale as is used for the diagram of the truss. We will assume that the train is headed to the left hand. Commencing at the left hand, write over each wheel the summation of the weights up to that point ; on another line above write the moments of all the wheels to the left of the particular wheel about that wheel.

The first line will give the weights P and W for any position of the train, so the placing the diagram to obtain $W - n\,P = $ a minimum, or $W\,y = P\,L$ is readily done.

The second line above will in like manner give the moments $(m\,P)$ or $(m\,W)$ or the moments of the loads about a, or the end of the span. Where the last moment does not correspond to the end of the span, it must be increased by $W \times d$, the distance from the last wheel to the end of the span.

The maximum shear at (a) will then be: $\dfrac{(m\,W) - n\,(m\,P)}{L}$

The maximum moment at (a) (different position of train) will be—

$$\frac{(m\,W)}{L}y - (m\,P)$$

An absolutely mathematically correct result can thus be obtained in a very few minutes, after a little practice. The preparation of a diagram for any engine and train load is only the work of a few minutes' time.

The method has therefore the merit of being no longer than that by uniform loads, and is far more exact, for it gives perfect accuracy for all positions of the loads, without any of those allowances for excess, etc., which are needed in the old methods.

This method has never been published before, but is in general use among the American bridge companies, by gradual transfers. The method cannot be applied to multiple system trusses, but it can be to girders with a continuous web.

The advocates of the method by uniform loads make the claim that from the rapid increase of the loads upon our cars we are fast approaching the time when the loads on the wheels of our cars will equal the loads on our engine wheels, totally forgetting that the distance between the wheels is as important an element as the weights. As long as the engine drags the train by the adhesion of its wheels, its weights and their concentration on a short wheel base must produce effects that will always exceed those produced by the car loads.

The problem before us, therefore, is not the selection of some maximum uniform load which will give us imperfectly proportioned structures; but how far can the designers of engines go in their loads and spacings of the driving wheels? Is there any maximum, or can they go on indefinitely? The want of harmony or rather the independence of the different departments of our railroad organization, has left this important question to be solved by the tentative system, rather than by any scientific study or investigation. That all of the roads in our vast system should adopt one maximum train load for designing their structures, even could such a maximum be decided upon, cannot be expected. The peculiarities of each road as to grade, alignment, traffic and financial strength must lead to variations.

That every road should build the best and strongest bridges that a sound and far sighted financial management will admit, is true. But we cannot expect more than this. Bridges, therefore, will continue to be built of various capacities. Our duty as engineers, is to strive to advance their capacity as far ahead of present needs as the funds available will permit.

We may accomplish this purpose best by bearing in mind certain principles:

First.—By selecting for our live loads typical train loads, where the engine loads will bear a reasonable relation to the following train load. The growth of the traffic and development of the train loads will then be more likely to increase the strains on all parts of our structures proportionately, and they, therefore, can be used till they have reached their maximum justifiable duty; instead of removing them, as is now so frequently the case, because the floors, counters or detail parts are overworked.

A recently constructed bridge is known to the author, which was designed for a very heavy load, so badly arranged that parts of the structure are only of a capacity equal to the present service of the road, while other parts have double this capacity. The material in this bridge, properly arranged, would have given a structure at least fifty per cent. stronger for future traffic, than it can now do.

Second.—By not increasing our unit strains in proportion to our faith in new forms and new material, where necessity does not compel us.

To-day we are able to get some steel forms at a less price than iron ones, and some others at the same price. Before long we will get all steel cheaper than iron. Why not take advantage, when we can, of this advance towards a stronger material to increase the capacity of our bridges? Why endeavor to push the strains up to the limits of our faith in its capacity for the various forms?

Camber.—The desired camber is obtained by making the length of the top chord panels to exceed those of the lower chord by an amount slightly in excess of the sum of the compression and elongation of the top and bottom chord panels respectively, under the working strains.

For strains in the top chord of 8 000 lbs. per square inch and 10 000 lbs. for the tension of the lower chord and a modulus of elasticity of 24 000 000 lbs., the sum of the changes in the two chords would be

$$\frac{18\ 000}{24\ 000\ 000} = \frac{1}{1\ 333}.$$

This amount of increase would, however, only provide for the deflection due to the chords. To provide for the additional deflection due to the web system of the trusses, and to allow some excess, this amount is usually increased one-third, which makes $\frac{1}{1\ 000}$, which corresponds very closely to "$\frac{1}{8}$ inch for every 10 feet of panel."

This becomes a very convenient rule for the shops where English measures are employed. It is entirely independent of the depth of the truss and only requires modification when the unit strains differ from the above averages.

Lateral and Transverse Bracings.—The general practice in America is to brace our trusses for a definite amount of lateral force per lineal foot of the structure, instead of assigning so many pounds of wind pressure per square foot of exposed surface.

This has arisen from two causes: 1st. The indefiniteness of the methods of estimating the exposed area, thus giving unfair advantages to the less scrupulous competitor. 2d. The recognition by practical engineers that there are other forces requiring lateral resistance, even in districts that may be windless.

The rule formulated for the Erie specifications in 1878 and now generally adopted, was as follows:

"To provide for wind strains and vibrations, the top lateral bracing in deck bridges and the bottom lateral bracing in through bridges shall be proportioned to resist a lateral force of 450 pounds for each foot of the span, 300 pounds of this to be treated as a moving load.

"The bottom lateral bracing in deck bridges and the top lateral bracing in through bridges shall be proportional to resist a lateral force of 150 pounds for each foot of the span."

"In no case shall any lateral or diagonal rod have a less area than ½ of a square inch."

While the above rules correspond approximately to 30 pounds pressure on the projected surface of a train of cars and the trusses, it was only selected after comparing the results with existing bridges up to 200 feet spans, which gave satisfactory lateral action under rapidly moving trains.

5. Strength of Material and Parts of Skeleton Structures.

In the early days of iron bridge building, the knowledge of the strength of materials was very limited. The early experiments of Fairbairn and Hodgkinson comprised about the extent of our knowledge of the strength of iron.

Crude tests made upon grooved specimens of a small size, lead to much misconception of the capabilities of American bar iron. In a pamphlet by Mr. Wendell Bowman on his Harper's Ferry Bridge, built in 1851–52, he states

"the tensile strength of the best American bar iron tables as 80 000 pounds per square inch. Its practical value is generally rated at about one-fourth the nominal value. In the diagram (*his strain sheet*) the highest value of iron is 16 000 pounds, being reduced below any probable rate of fibral separation in any previous data."

Even as late as 1870, Captain Eads was assured by the manufacturers of bar iron that there was no difficulty in furnishing bars of any size, capable of standing 60 000 pounds per square inch, and he made the contract for the St. Louis Bridge with this expectation, only to be disappointed.

In 1878, when the requirements for the strength of bar iron as contained in the Erie specifications, making a reduction in the ultimate strength as the size of the bars increased, was submitted to the most experienced iron masters for their criticism, they were condemned by all with but one exception, Mr. Andrew Kloman, of Pittsburgh: 1st, as being entirely too low in tensile requirement for good bar iron; and 2d, in making any allowance for increased section of the bars; "there being no reason why the fibers in large bars should not be as strong as in the smaller."

It may be unnecessary to state, that when these specifications were enforced, and the acceptance of the material became dependent upon the test made on either the bars themselves or specimens cut from the same, it was found that only the best and highest priced bar iron could meet the requirements.

The general tendency, since then, has been to relax them somewhat for a broader competition.

The Phœnix Iron Company and the Keystone Bridge Company had crude testing machines at their shops previous to 1870.

They were used to test the strength of eye-bars and wrought-iron columns. In some of the more important structures, all the eye-bars were tested up to a strain of 20 000 pounds per square inch.

While these machines were crude, they did serve to develop the detail proportions of the eye-bar and column used by these companies. The use of steel and new forms of members in the St. Louis Bridge, in connection with the importance of the structure and the yet untried method of erecting as a cantilever, impelled the engineer and the contractors to unusual efforts to determine the capacity of the material and of the forms to be used.

Of the many thousand tests made during the construction of this work, but a limited number have ever been published; the general deductions, however, soon passed into accepted doctrine, and developed a keener desire for a more refined knowledge of the influence of form and proportion upon all our bridge members.

The eye-bar had been well developed under the crude tests previously made by the early bridge companies.

Full sized tests had also been made on a limited number of wrought-iron columns. But no marked progress was made in the forms of compression members, till after the tests of Mr. Bouscaren on the forms then in use. His tests not only showed the inapplicability of Gordon's formula, but also the possibility of much better results by improving the detail of the open and box columns; the only forms of columns without a proprietary claim and the ones best adapted to the American style of bridges; as they gave ready facilities for connections without the use of cast-iron.

From this time forward rapid progress has been made in the detail and proportion of all our full sized bridge members.

To-day every first-class bridge manufactory has its complements of testing machines, to test with all the refinements either samples of the material or full sized members in compression, tension or transverse strain. Our knowledge of the strength and capabilities of our material and of the usual forms employed in the American style of bridge is such, that no first-class bridge company in America hesitates to accept the clause now general in all specifications, that "full sized members may be tested to destruction," with the sole proviso that the expense of testing and cost of the piece shall be paid for by the purchaser if it satisfies the requirements of the usual specifications.

This positive knowledge of the capacity of our full sized members marks one of the great advantages of our system of bridges over all others. Our "factor of ignorance" has been reduced to this extent; a no mean portion.

We, therefore, have a right to claim, that as our working strains are as low and in many cases lower than those used in Europe, with our more perfect knowledge of the strength of our members, we have in our first-class structures a greater factor of security than prevails in European bridges.

In the appendix will be found tables of the abstracts of the more recent tests on eye-bars and compression members.

6. MANUFACTURE OF BRIDGES.

The great demand for iron bridges in a country as vast as the United States caused the formation of companies for the special purpose of manufacturing and erecting bridge structures. They established special plant and created a corps of engineers, who have made this branch of engineering a special one.

There are to-day in America more than forty bridge building companies manufacturing railroad and highway bridges.

Of these, at least a dozen are capable of constructing bridges of every size in a first-class manner.

The shops, which can be called especially shops capable of constructing the largest class of railroad or highway bridges, are capable of turning out by the year 125 000 tons of bridge work.

It would be safe to estimate that all the shops could turn out 200 000 tons, or approximately 80 miles of 100 feet spans of single track railroad bridges, per year, if their plant were devoted exclusively to this work. Other iron work, as roofs, iron buildings, piers, elevated railroads, etc., however, make a considerable figure in their yearly output. Some few of the bridge shops have been constructed to do riveted work almost exclusively.

The typical American bridge shops are, however, fitted to do any class of bridge, girder or roof work, whether it be exclusively riveted, or combined riveted and pin-connected work. Each company has, therefore, the following arrangements for receiving the iron, and putting it through all the processes to the completion ready for shipment:

First.—Receiving yard, where the iron for each bridge is properly classified and stored.

Second.—Department for straightening, where pieces can be straightened with more accuracy than can be obtained directly from the rolling mills.

Third.—Template and pattern shop, for preparing the templates for the rivet and pin holes and crude shapes and dimensions of all pieces, with the proper allowance for final tool finish.

Fourth.—Laying out shops, where each individual piece of iron is carefully marked in accordance with the templates.

Fifth.—Punch and shear shop, where all the iron is punched and sheared.

Sixth. —Fitting up shop, where all the iron for riveted members is assembled and bolted together ready for the riveting machines.

Seventh.—Riveting shop, with its proper complement of air, steam and hydraulic riveting machines.

Eighth.—Machine shop for planing, boring, turning, etc., to complete the finished bearing surfaces of all the members.

Ninth.—The upsetting and forge shops for making eye-bars and all forged parts required.

Tenth.—Painting and shipping sheds and yard.

The aim in the construction is to pass the material from the time of its receipt from the rolling mill to its final shipment through the necessary steps with as little waste labor in handling as possible and to perform all work by machinery in preference to hand labor.

While the laying out and riveting are done with all care and accuracy, the lengths of the members and the fitting of the same together do not depend upon the accuracy or neglect of these processes.

The length of all abutting members and distances between centers of pin holes are determined finally at the machines for planing the abutting ends or for boring the pin holes.

Machines for operating on each end of such members are provided with iron beds and extremely accurate methods of setting the same to any required distances.

Such machines once set and operated with proper precautions in regard to uniform temperature guarantee great accuracy of duplication of all similar parts.

It is sometimes claimed that there is an error in lengths of parts due to the usual allowances for play of the pins in the pin holes, varying from $\frac{1}{64}$ to $\frac{1}{32}$ of an inch, according to the size of the pins. This is an error; this allowance is provided for by either measuring from out to out of the pin holes for tension members and from the inside of the pin holes for compression members, or by taking it into consideration where the lengths are given from center to center.

The surety of the fitting of the members of even the largest structures after they have passed through a properly organized bridge shop is such that no assembling of the finished members of a structure is ever made at the shops, except in extremely crooked and complicated structures, and even then not as a whole, but only sufficient to test the fitting of specially intricate connections.

The rapidity of the erection of our structures and the satisfactory manner in which they come together in the field without any tool work prove the certainty of the American methods.

A pin connected truss is composed of the following class of members:

1. Top chord sections.
2. End posts.
3. Intermediate posts.
4. Pedestals.
5. Lower chord bars.
6. Diagonal web bars.
7. Pins.

The top chord sections and end posts are usually similar in form. They are usually formed of two or more vertical channels, either of rolled forms, or built of angles and plates, connected on top with a cover plate and on the bottom with lattice bars.

In the best practice most of the material for these members is concentrated in the vertical channels; the top plate besides acting as a cover and stiffener during transportation, gives an unbalanced section sufficient to counteract the tendency to deflect due to the weight of the piece. The pin holes are reinforced by additional plates to reduce the bearing pressure to the allowed limits. The amount of rivets in these pin plates and at the ends of the sections is made large enough to distribute the localized pin pressure over the section of the whole chord.

The chords are spliced at one side of the pin hole to insure a full pin hole for more convenient boring and to enable the rivets in the splice to be driven after the parts are assembled.

After these chords are riveted together in the shop the abutting ends are carefully tooled to square surfaces and exact lengths and the pin holes bored. The accuracy of the finished lengths and diameter of pin holes and the character of the tool work having been passed upon by the inspector, the piece is marked, painted and stored for shipment.

All posts are made with pin bearings only. The intermediate posts are usually made of the open form, consisting of two channels, either rolled or built up of plates and angles, latticed on both sides. The sizes of the end or batten plates and the lattice bars have been evolved from experience in transportation and from the study of the results of full sized tests.

The ends of the posts are usually forked (see Fig. R, Plate XXVII), in order to pack between them the tension bars to produce a more compact joint.

These ends are stiffened by thickening with plates of sufficient length to make the compressive strength of these ends fully up to the capacity of other parts of the same post.

The pedestals are riveted members formed of plates and angles. They are proportioned for the proper bearing pressure and with rivets sufficient to meet all the strains. Their base is determined by the allowed pressure upon the masonry.

Under the pedestals of one end of the span, nests of friction rollers proportioned according to an accepted rule are placed, to provide for the expansion and contraction of the trusses.

Lower chord bars.—These are flat bars with forged eyes at each end. These bars have, in the past, been made by various methods. But economy in their manufacture has developed special upsetting and die forging machines of great power, using either steam or hydraulic pressure, by which they can be made with great certainty and at a very moderate cost. Only by such perfected machinery and great experience in the working could such accurate results as are demanded by our ordinary specifications be attained. The center of the heads must be in the center line of the bar and the heads symmetrical to this line. The heads must be clean, smooth and of proper thickness. The necks must be of proper shape and upset so as to be full in all dimensions. The lengths between centers of the heads must be very close to correct length or when bored there will not be the proper excess for the eye. To produce such bars economically all reheating and reworking to correct errors must be avoided.

The process, of course, requires the best material, whether of iron or steel.

All steel bars are annealed after they have been forged.

Our summary of some recent tests made in the ordinary course of the execution of contracts shows how perfectly these bars answer to the full capacity of the material.

Web-bars.—These are similar to the lower chord bars, except the counter-rods, which are made in two pieces and connected by sleeve nuts or turnbuckles. It is the best practice to secure these turnbuckles or sleeve nuts firmly in their position after these rods have been properly adjusted.

Pins.—These are made of round iron or steel, turned perfectly smooth, and to accurate diameters, with an allowance of $\frac{1}{50}$ inch for pins under $4\frac{1}{2}$ inches diameter, and $\frac{1}{32}$ inch for larger pins, less diameter than the bore of the pin holes. These pins have wrought-iron nuts at each end, which are tightly screwed up and secured after the bridge is connected together.

The Iron Floor System of our bridges consists of cross floor beams at each panel point, with a pair of stringers under each track. These floor beams and stringers are usually plate girders. This system is connected to the trusses usually in one of two ways; either by suspending from the pins or by riveting to the vertical posts for through bridges; or for Deck Bridges, either resting on the top chords at the panel points, or by riveting to the posts. Each particular case must be worked out to accomplish the purpose best, considering the local requirements of clearance for floods or navigation, head room, etc.

Lateral and Sway Bracing. — The horizontal and vertical bracing between the trusses have generally in the past been similar to the other characteristics of our type of bridge, rectangular systems of tension rods and compression struts connecting to the pins of the main trusses. Of late years the preference for more rigid forms for the parts of these comparatively light trusses has led to the use of angle iron bracing instead of tension rods.

7. Erection of Bridges.

The great risk involved in erecting important structures over rivers subject to sudden floods, ice jams and similar dangers, emphasizes the importance of having a style of structure which can be rapidly and surely assembled without detriment to the perfection of the workmanship or accuracy of the connections.

The American pin-connected bridge is especially suited for this purpose, the connection of the comparatively few compact members being readily completed after they have been hoisted into position by the driving into place the connecting pin and securing the same by setting up the nuts.

The only rivets to be driven in the main trusses are usually those in the plates connecting the top chord and end post sections. But these do not in any manner affect the perfect action of these members, which depends entirely upon the machine surfaces of the abutting ends and the turned pins. They serve solely to insure the bearing surfaces from side displacements through any accidental blows. They can be connected by bolts, or the rivets may be driven at any time, even after the bridge is in use for trains.

The rivets in the floor system can likewise be driven after the bridge is free from the false work.

The erection, therefore, of a pin-connected bridge is very readily performed, and with men experienced in this class of work with marvelous rapidity.

The importance of the erection with rapidity compels a careful design for the detail of all connections, and also often gives weight in favor of a special style over all others.

The additional security of the cantilever method of erection favors that form of bridge for long spans over deep gorges or rivers where the risks are unusually great.

The selection of a cantilever style of bridge for spans less than 500 feet is entirely unjustified, except where the special merit of this method of erection outweighs the objectionable features of the system— excessive deflection and reversal of strains.

The rapidity with which our pin-connected trusses can be assembled and swung clear of the supporting false works has been so often demonstrated, it would be useless to enumerate examples.

A 250-foot span railroad bridge has been erected in sixteen working hours, taking the material from the storage yard 1 000 feet from the bridge, without any emergency demanding unusual exertions.

It would be no exaggeration to assert that any span up to 250 feet could be erected within the limits of one day-light, so as to be self-sustaining and independent of all risks from loss of the false works.

The erection of the two channel spans of the Cairo bridge is an example of rapid erection, which illustrates the possibilities of the American system of construction.

These spans were each 518 feet 6 inches center to center of end pins; 61 feet deep center to center; 25 feet wide center to center of trusses; panel lengths 30 feet 5¾ inches. Total weight of one span, 2 055 200 pounds.

The first span was erected in six days. After this span was erected, the false works were taken down, the supporting piles drawn and redriven for the second span, the false works again put up and the second span erected.

The whole time covering the erection of the two spans and moving the false works was one month and three days, including five days lost time, waiting for the completion of the masonry.

The false works and traveler were in position ready to commence the erection of the second span by 2.30 p. m., October 30th, 1888. At 2.50 p.m., November 3d, the trusses of this span and the top bracing were all connected. No work was done at night. The material was run on trucks about 1 025 feet from the storage yard to the nearest end of the span being erected. About twenty-four men were employed in delivering the material and fifty in erecting and connecting together.

The false works stand about 104 above low water. The bents being 72 feet high above the capping of the piles. The depth of water at its low stage is about 20 feet. The piles were from 50 to 75 feet long.

The erection of these spans was done by William Baird & Co., sub-

contractors, under the Union Bridge Company, for this part of the
work. Plates XX to XXIV are reproductions of photographs taken
every twenty-four hours to show the daily progress of the erection
of this second span.

8. TYPICAL AMERICAN RAILROAD BRIDGES AND THEIR RELATIVE MERITS.

While the pin-connected bridge is recognized as the one which
can be especially called the American type, American bridge engineers have not failed to "search all things and hold fast that which
is good," not only within our own experience but within that of
European practice. We have had one great advantage over our
European brothers, that their works and practice are so much
prompter and fuller given in the technical literature of the day.
We have had another advantage from our allying the several branches
of designing, manufacturing and erecting more intimately than is usual
in European countries. We have been less liable, therefore, to give
undue weight to the claims of theorists or the riders of hobbies. Nearly
every imaginable form of structure has been tried in the past by some
enthusiast.

After an experience unequaled for its variety and extent, we are ar-
riving at a very general uniformity in the styles of bridges to be adopted
for different spans.

Ten years ago, of a thousand bridges of which no two would be
alike, a practical bridge engineer could almost invariably determine the
designer and manufacturer, either by the general style of the bridge or
by the use of some peculiar forms of member or detail.

To-day the use of a peculiar form of truss, member or detail is due
more to the special necessities of the case than to the idiosyncrasies of
any particular individuals.

While no two bridges to-day are exactly alike, there is a general
approximation to certain broad types in styles and detail.

Plate Girders.—These are very generally used for spans up to 65
feet, or lengths which do not require more than two ordinary flat cars 33
feet long to transport from the shops to the bridge site.

Some railroad companies have extended their maximum length for
plate girders to three car lengths, or about 100 feet.

These girders are riveted completely at the shops. Plate girders
completed at the shops in this manner give excellent satisfaction. For

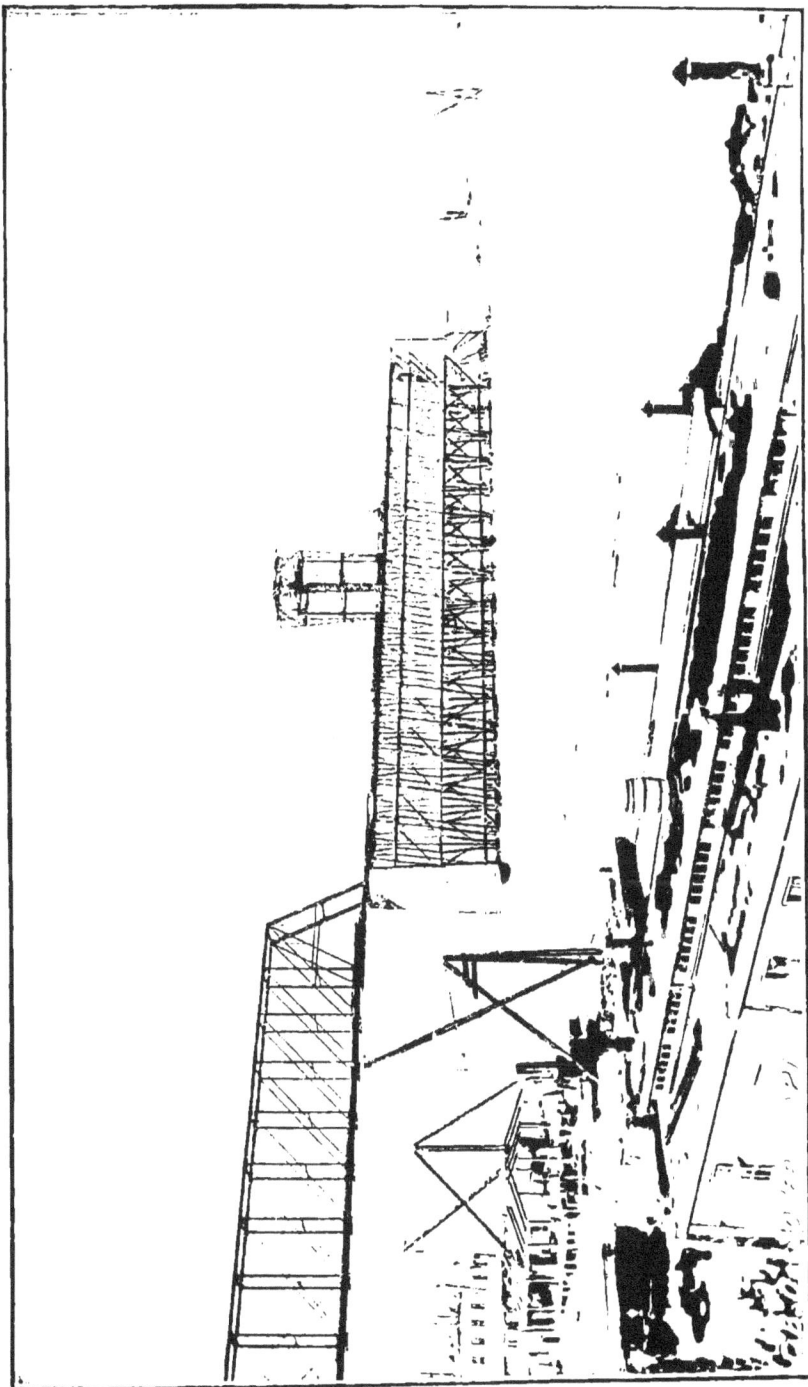

PLATE XX.   CAIRO BRIDGE.   Vol. XVI. p. 12.

Plate XXI.

CAIRO BRIDGE.

Vol. XXI. p. 12.

PLATE XXII.   CAIRO BRIDGE.   Vol. XXI. p. 42.

CAIRO BRIDGE.

Plate XVIII.

Vol. XXI, p. 42.

equivalent strengths, they are cheaper up to at least 65 feet spans than lattice girders. Their maintenance is also less. Their relative security is far greater ; one per cent. of faulty rivets will make a much greater reduction of strength in a lattice girder than in a plate girder to do the same duty. Plate girders are cleaner than lattice girders, being free from the numerous recesses and corners peculiar to this latter type, and hence less exposed to the accumulation of dirt, moisture and their accompanying oxidation.

Lattice Girders.—Riveted lattice girders have been used quite generally on all our railroads for short spans and on certain lines of railroad for all spans. For the shorter spans, as before stated, preference is now strongly in favor of plate girders. For the larger spans the necessity of performing so much riveting of important connections at the bridge site, where that care and accuracy attainable at the shops cannot be depended upon, does not render them acceptable to many engineers. And the necessarily increased risks and cost of the erection lead most manufacturers to prefer some other style of bridge, except for special cases.

While lattice bridges have given good satisfaction where they have been well proportioned, they have no merits which cannot be also had from our pin-connected types. They have not, therefore, made any progress for the longer spans.

On most of our roads they are now generally limited to the spans between the longest plate girders and the minimum pin-connected spans; a limit not definitely fixed, the lower limit being determined by the relative cost of plate and lattice girders and the ability of the roads to transport long plate girders ; and the upper limit by the personal preference of the engineer. These limits may be broadly stated as somewhere between 60 and 120 feet. The riveted form of connection has an advantage over the pin form for the lower spans, to be determined by the circumstances of each case, from the additional stiffness of the connections and the form of the members. This is partially attained by some designers by using stiff members for both tension and compression in their short pin-connected trusses. It would be possible to make an equally stiff pin-connected bridge for these lower spans, and at the same time attain that perfection of fitting and lengths of parts peculiar to this form of truss, if those connections, at which reversal of strains occur, were also rigidly bolted or riveted together. This might be objectionable to the theorists who object to secondary strains. The author does not consider these, if restrained within proper bounds, as so objectionable ; pin-connected trusses cannot be considered as free from them, as long as there exists such a thing as friction.

Pin-connected Spans.—With the exception of not over a few hundred spans, the longest of which are 260 feet (lattice bridges), the iron railroad bridges of the United States for spans over 100 feet, amounting to about 7 000 spans, aggregating 210 miles in length, are of the pin-connecting type.

The American people are of too practical a character to have built this quantity of bridges of a peculiar type, with the knowledge of other forms in use, both here and in European countries, and after thirty years' experience in their use, to still adhere to them, if their merits were not strong and positive ones.

Through all the changes of styles of trusses and forms of parts, the connection of the members by means of pins, and the attaining accuracy of lengths and fittings by machining the joints, has been persistent.

Forms of trusses with more than a single system of triangulation have gradually been rejected, and have now disappeared from first-class bridge designing.

Such subdivisions of the trusses as will give the greatest concentration of parts, with long panels, and without any theoretical or practical ambiguity of strain under a miscellaneous series of wheel loads, are now only adopted.

Forms of members requiring cast-iron joint boxes or other uses of this material have become obsolete.

The old forms, like the Bollman, Fink, Lowthorp and Post trusses (see Plates XII, XIII, XIV, XV, XXV, XXVI), have disappeared from American practice. The double intersection Whipple or Linville is rapidly following them. Generally bridges are made with parallel chords and equal panels; this gives the minimum number of different lengths of parts, which leads to greater economy and accuracy of manufacturing. For the longer spans, the depth may be reduced with advantage at the end panels.

The longest pin-connected truss span in existence is the recently completed channel span of the Ohio River Bridge at Cincinnati, 545 feet from center to center of end pins, 84 feet deep at the center and 60 feet at the end posts; panels, 27 feet 1½ inches; trusses, 30 feet apart, center to center. It carries a double-track railroad between the trusses, and has on each side a wagon-way and foot-walk, 16 feet wide.

The persistence of the pin-connected type of structure, now that the bridge engineer acting for the railroad companies has become an important factor in the problem of bridge construction, shows that it is not solely due to the preferences of the manufacturers; but that the operation and maintenance of such structures are also in favor of this type.

The typical American railroad bridge is a skeleton structure, pin-connected at all the principal articulations. Its essential characteristics in addition to being connected by pins are :

First.—So formed as to reduce all ambiguity of strains to a minimum ;

Second.—Concentration of parts ;

Third.—Facility of manufacture ;

Fourth.—Perfection of lengths and fitting of all the members, so as to reduce to a minimum all riveting or mechanical work in the field ;

Fifth.—Readiness with which the individual members can be assembled during erection.

9. Amount and Kind of Bridges on the Railroads of the United States.

The author attempted to collate from the official reports of the State Railroad Commissioners the data from which to form an estimate of the amount and kinds of bridges on our railroads.

He found that many States had no bureaus which collected this information. Also, that many reports were indefinite and evidently incorrect.

In order, therefore, to obtain more definite and reliable information, and also to get data which could be considered as truly representative of all the geographical divisions of the country, he sought it directly from the principal large systems of railroads. Many of these systems promptly placed at his disposal very full and reliable information as to their bridges.

Taking from the Railroad Commissioners' reports such data as appeared correct and complete, and adding thereto this additional information, avoiding as far as possible any duplication of the same roads, data covering nearly 60 000 miles of railroad, fairly distributed over the whole of the country, were obtained.

In order to make these data as nearly comparable as possible, the length of all double-track bridges was reduced to the corresponding length in single-track bridges.

In like manner, the length of all main tracks, omitting sidings and turnouts, was put into terms of single track.

Elevated railroads, which are composed mostly of bridges or trestles, have been omitted entirely.

Table No. 3 gives the general data as to quantity of bridges and trestles and the average rate per mile of track.

It shows that the relative amount of bridges and trestles varies in different districts from 58 feet per mile to 231 feet per mile. This last, however, is excessive, from including the crossing of Lake Pontchartrain, near New Orleans, on a trestle 22 miles long. Omitting this, we would get only 162 feet per mile as the maximum.

These variations are not entirely due to geographical location, as might appear at first thought. They are also affected by principles governing the original location of each road or division of a system. The alignment and grade may have been sacrificed to the avoidance of bridges or trestles, or the contrary.

From the large mileage covered by our table, we can rely with considerable confidence upon our average.

TABLE No. 3.

System of Railroad, or State.	Miles of Road.	Total length of Bridges and Trestles in feet.	Lineal feet of Bridges and Trestles per Mile of Road.
New York Central and West Shore Railroads..........	2 894	364 722	126
New York, Lake Erie and Western Railroad...........	1 514	95 509	63
Other Roads in New York....................	3 586	445 900	130
Roads in Pennsylvania..............................	4 352	336 957	77
Roads in New England................................	2 199	176 700	80
Wabash System......................................	1 636	160 025	98
Missouri Pacific System..............................	4 707	566 953	120
Chicago, Milwaukee and St. Paul Railroad............	5 727	614 736	107
St. Louis and San Francisco Railway..................	1 441	130 075	90
Denver and Rio Grande Railroad........	1 458	102 195	70
Union Pacific Railroad................................	4 754	276 032	68
Louisville and Nashville Railroad.....................	2 495	322 679	123
Queen and Crescent System..........	1 139	299 222*	231*
Roads in Illinois.....................................	8 539	707 535	83
Roads in Michigan.................	4 151	249 345	60
Roads in Iowa....................	7 778	1 049 386	135
Central Railroad and Banking Company of Georgia....	1 487	173 975	117
Totals	59 857	6 071 946	101

* Included the crossing of Lake Pontchartrain, a trestle 22 miles long.

Taking, therefore, 100 feet per mile as our basis of estimate, we have for the 160 000 miles of railroad in the United States, 16 000 000 feet or 3 030 miles of bridges and trestles.

TABLE No. 4.

Miles of Railroad.	Trestles and Spans under 20 feet.	Spans from 20-50 feet.	Spans from 50-100 feet.	Spans from 100-150 feet.	Spans from 150-200 feet.	Spans from 200-300 feet.	Spans from 300-400 feet.	Spans from 400-500 feet.	Spans over 500 feet.	Total.	Average per Mile of Road.
26 228	2 299 758	83 181	94 165	149 121	80 551	29 542	5 677	1 211	1 040	2 746 246	104.7 feet.

Table No. 4 gives the distribution of the bridges upon 26 000 miles of railroad into spans of different lengths.

Using this as a basis of estimate, the 3 030 miles of bridges and trestles in the United States should be distributed as follows:

Trestles and spans under 20 feet......... 2 424 miles = 727 200 spans.
Spans from 20–50 feet................... 121 " = 18 150 "
" 50–100 " .... 130 " = 9 100 "
" 100–150 " .... 190 " = 8 000 "
" 150–200 " .... 109 " = 3 300 "
" over 200 feet..................... 56 " = 1 150 "

3 030 " = 766 900 "

The above includes all bridges of either wood or iron.

Using the detailed information in the author's possession as to the bridges on the above 26 000 miles of railroad, which are of wood, combination and iron, we obtain the following estimate for the amount of iron bridges on our railroad system:

Iron spans under 20 feet....... 17 miles = 5 100 spans.
" 20–50 feet.......... 86 " = 12 900 "
" 50–100 " .... 66 " = 4 600 "
" 100–150 " .... 93 " = 3 900 "
" 150–200 " .... 69 " = 2 100 "
" over 200 feet........ 49 " = 950 "

Total........................ 380 " = 29 550 "

The author's own office records contain the following iron bridges over 200-foot spans, and they are very incomplete for spans less than 300 feet:

Spans over 500 feet......................... 2.5 miles.
" 400–500 " .... 3.9 "
" 300–400 " .... 5.0 "
" 200–300 " .... 15.0 "
Total............................... 26.4 "

From the previous figures it would appear that there are now in existence on our railroads the following wooden and combination trestles and bridges:

Trestles and spans under 20 feet............. 2 407 miles.
" 20–50 " .... 35 "
" 50–100 " .... 64 "
" 100–150 " .... 97 "
" 150–200 " .... 40 "
" over 200 " .... 7 "
Total............................... 2 650 "

Of the 2 400 miles of wooden trestle, we can consider one-quarter as only temporary, to be filled in as embankment. Of the remaining 1 800 miles at least 800 miles will be maintained in wood. This leaves 1 000 miles to be replaced gradually with iron bridges and trestles of such spans as may be most suitable for each location, probably from 50 to 200 feet.

This would ultimately make 1 600 miles of bridges for the 160 000 miles of railroad now in existence, of which only 380 miles are in iron at the present time.

The substitution of iron for the existing wooden bridges will, of course, be a gradual one. The amount of work involved in this, with that required in building iron bridges for new roads, additional crossings of our large rivers to connect present and future systems of railroad, and the demands for city and highway bridges, will furnish plenty of employment for our bridge building firms.

The author has no reliable basis for estimating the amount of highway bridges demanded by the present or the future. A comparison of the relative crossings of streams by railroads and highways in the more developed parts of our country, leads him, however, to the belief, that the lineal feet of highway bridges will be four to five times that required for railroads only.

10. FAILURE OF BRIDGES.

In consideration of the facts already presented, it need not bo a matter of surprise, that bridges have failed under train service in America.

The data in regard to the number of such failures have been magnified and distorted.

That such distorted evidence should have been submitted to a foreign society, without any opportunity for those acquainted with the truth to counteract the false impressions and conclusions, is to be regretted.

In the paper referred to, 251 bridges are given as the number of failures in the last ten years, classified as follows, by the author of the paper: 57 knocked down by derailed trains; 30 "square falls"; 96 uncertain; 5 occurred during replacement or repairs, and 63 unclassified.

That the great majority of these bridges were old and defective wooden bridges is not mentioned.

In addition to this generalization, there are given a number of photographs with incomplete statements of the failures they represent.

Of the nine railroad bridges, in regard to which their identity or any reliable statements are given, we are able to summarize as follows, with the aid of other facts in our possession:

Three knocked down through derailment of the train, one from a broken axle, one by a cow on the track, and one from neglect to spike the newly laid rails.

One split open by two trains colliding within the trusses; this was an old-fashioned structure with cast iron top chords.

One knocked down by a steam derrick too large to pass through the bridge.

One wood and iron combination bridge, seven years old, which failed in its wooden members under a load far in excess of its designed load.

One, the Bussey Bridge, an abortion in design and construction, and a bridge in which no engineer had any part.

One which is stated to have been culpably overloaded, and to have withstood this excessive overloading for a long period without showing any apparent signs of distress.

One showing a lattice bridge distorted by a train collision and which still carried the train.

The above is a summary of the evidence presented to show that

American pin-connected bridges are not so safe and reliable as those of a lattice form. It is fair to presume that this is the strongest evidence that the facts would admit.

The same authority, quoted by the author of the paper referred to, gives more recently a fuller account of the bridge failures for the ten years ending January 1st, 1889; total number, 265, of which only 38 are known to be of iron.

In his total he includes 27 wooden bridges burned, 39 carried away by floods, 8 failing during repairs, 60 knocked down, 34 square falls, 97 unknown.

Of the 34 square falls, 10 only are of iron.

The one iron bridge recorded during 1888 as a square fall was a through plate girder, failing from the breaking of a plate carrying the cross floor beams. More detailed information of the other nine is not given, so we are unable to determine the kinds of iron bridges and the causes of the failures.

It may be necessary to state that these reports of accidents are not based entirely on official reports; but are collated from the newspapers, and are undoubtedly full of sensation and inaccuracies.

Does any one advocate the designing and building of bridges to withstand the impact of a railroad train, or the bursting effect of piling two trains on one another inside of the trusses? Are such accidents to be classed as bridge failures or as failures of management?

That bridges of the lattice type have given good results in many cases of derailment, etc., proves nothing. Scores of similar cases, showing that pin-connected bridges do stand much abuse, and are as capable of resisting occasionally these extraordinary strains as any other kind of bridge, could be given by every practical bridge engineer in America.

They prove nothing in regard to the merit of any particular form of truss. They simply point to the need of better means to prevent and remedy cases of derailment.

It is a very gratuitous claim to assume that the bridge accidents on railroads in America are in any manner due to the accepted forms of our bridges, or that other forms of bridges would have given different results under the same circumstances.

The number of bridge accidents on American railroads, due to causes or forces which a bridge should be designed to meet or resist, is an exceedingly small one, when compared, as it should be, with the total number of bridges on our railroads, and when the circumstances under which many of them have been built be given proper consideration.

The total number of spans of railroad bridges in the United States over 50 feet span, is 21 550, of which number 11 550 are of iron.

Due weight must be given to the rapid development of our railroad system, and the want of appreciation in the past of the demands for heavier rolling stock and correspondingly heavier bridge structures.

The American idea of building cheap railroads far in advance of the immediate demands of the regions through which they run, to settle those districts and build up a future paying traffic, has compelled the use of cheap and light bridge structures.

Even on the roads in the older portions of our country, the earlier bridges were built for loads far below present necessities.

Economy has also, in many cases, favored the lowest bidder, regardless of the incapacity or unscrupulousness of the bidder. There have been and still are many bridges of an inadequate strength in use on American railroads.

They are disappearing very rapidly, especially upon our better managed and first-class railroads.

Under these circumstances, it is a great credit to our style of bridge construction, that so few, comparatively, of our bridges have failed.

The writer does not know of a single case of the failure of a modern all-wrought-iron American railroad bridge, which has failed under legitimate train service. He has had knowledge of many bridges which have for years done far beyond their intended duty with satisfaction.

That the bridges built in the past have been lighter than the demands of the future must not be entirely credited to the engineer. As a rule he has done fully as well as the means given him would admit.

The American engineer has reaped one great advantage from the defective and inadequate structures of the past, not only from those that have actually failed, but also those which have been worked beyond their designed capacity—a knowledge of the relative merits of the various forms and proportions of the members and their details.

These structures may be said to be full sized tests, from which much that is valuable has been learned.

We can feel that our knowledge of bridges has passed far beyond the study of strains and tests of specimens. We have not remained content with this limited field of knowledge, covering our ignorance under a so-called "factor of safety." We have determined the capacity and reliability of the full sized parts of our structures. We have also been enabled to study the working under excessive loadings, even up to rupture, of structures and their details which had outlived their intended purpose or which were inadequately designed from the beginning.

American bridge engineers and manufacturers have therefore legitimate grounds upon which to base their claims as to the great merits of our system of construction and proportioning.

The author desires to emphasize the statement that failures of bridges in the United States have no more bearing upon the relative merit of different systems of their construction, than the failures in other kinds of machines. The old-timed boiler explosions on our Western rivers did not convince any sensible engineer that cylindrical boilers were not suitable for high pressures. That machines inadequately proportioned for the demands which will be put upon them break down ultimately, especially when they fail to receive intelligent care and attention, is to be expected.

The intelligent investigator does not decide upon the merits of any developed system by the failures which are necessary steps in its development. Without variations and failures there would be no evolution or the survival of the fittest.

These variations and failures are interesting and instructive, but should not blind us to the merits of the final result.

The following remarks by Professor Unwin are worth careful consideration in this connection :

"If an engineer builds a structure which breaks, that is a mischief, but one of a limited and isolated kind, and the accident itself forces him to avoid a repetition of the blunder. But an engineer who from deficiency of scientific knowledge builds structures which don't break down, but which stand, and in which the material is clumsily wasted, commits blunders of a most insidious kind."

The author desires to draw attention to a series of articles on American and English bridges, by Mr. T. Claxton Fidler, in "Engineering," November 9th to December 28th, 1888. They are a valuable contribution, and are exceedingly fair in their treatment of the subject, considering that their author on some points has not the real facts of our best practice before him, and that he has mistakenly accepted a particular bridge company's specification as representing the general practice in our country.

He summarizes the following results as derived from the American system of competitive bridge construction :

"*First.*—It has developed much ingenuity of design, which has been directed, however, to the improvement of trusses in detail rather than to such broader features of design as are to be seen in the relative outlines of the Britannia Bridge, the Saltash Bridge and the Forth Bridge.

"*Second.*—It has attained to a great economy in details and the advantage obtained by the use of pin-connections, together with the attendant facility of erection has sometimes enabled the American maker to obtain the lead in the market.

"*Third.*—The system has necessitated and has produced an earnest inquiry into the theoretical principles of bridge construction, whose results have been embodied in certain specifications comprising a series of rules and regulations which are intended to insure the safety of the bridge.

"*Fourth.*—The defects and possible abuses of the competitive system are, however, acknowledged; and have been attended with numerous bridge failures.

"*Fifth.*—In quite recent times there has been a strenuous endeavor to repair the defective features of the system, and to obtain more ample guarantees for the safety of the bridge in the shape of more stringent rules; but so long as the rules are drawn with reference only to the real stresses, they must always be insufficient for the purpose in view, unless they are supplemented by the independent judgment of the engineer proceeding from merely arbitrary methods.

"*Sixth.*—In order that the rules may be effective for the purpose in view, they must be based upon a complete theory of safe construction, and not merely upon a theory of stresses; and must provide for all those requirements of stability and solidity which are instinctively recognized by the practical engineer, and which cannot be complied with by merely using a large factor of safety.

"*Seventh.*—If these requirements are to be formulated in mathematical shape, it would be necessary to enter upon that wider inquiry which has been referred to in these articles, and to specify a certain series of empirical and unreal distorting forces, which are to be assumed as acting at different points in the structure, and to be resisted by members possessing the strength which may be requisite for that purpose in addition to the strength which may be demanded by the real stresses. Such a theory of unreal stresses may perhaps be dispensed with in English practice as it has been hitherto; but there is no reason why it should not be framed upon common sense principles, and adapted to the requirements of the American competitive system, for whose purposes it seems to be urgently required."

It is true that our system of bridge building has not advanced in the direction of such designs as the Britannia and Saltash Bridges during the past. Whether it will be towards designs like the Forth Bridge in the future is very doubtful. The author believes that problems of this magnitude will be solved by the American bridge builder, when the opportunity occurs upon the broad principles of our past experience.

We are always ready to admire "big things," as we have each of these structures in its day, and to reap benefit from the valuable contributions which have been given us by their able engineers in the line of new investigations and experience.

We are ready to acknowledge the evils of the old competitive system, where "the independent judgment of the engineer," or "those requirements of stability and solidity which are instinctively recognized by the practical engineer," do not enter sufficiently to obtain a safe system of construction.

The most perfect system of rules to insure success must be interpreted upon the broad grounds of professional intelligence and common sense.

While our competitive system, guided by a general specification, is more apt to produce good bridges than dependence upon designs prepared by any one with narrow and limited practical knowledge, its best results follow when supplemented by the independent judgment of the practical bridge engineer.

APPENDIX.

ABSTRACT OF RECENT TESTS ON FULL SIZE BRIDGE MEMBERS.

TABLE No. 5.

Abstract of Tests on Wrought-iron Columns, Watertown Arsenal.—Symmetrical Sections—Pin Connected.

Set.	Test Number.	Ratios. $\frac{l}{d}$	Ratios. $\frac{l}{r}$	Area of Section. Square Inches.	Pin Bearing. Square Inches.	Diameter of Pin. Inches.	Ultimate Strength per Square Inch. Pounds.	Mode of Failure.
	17	10	27	12.	9.1	3½	35,070	Channels buckled near end of column.
	1109	10	27	12.2	9.1	:	33,420	Channels buckled near end of column.
	15	15	40	12.1	9.1	:	33,880	Channels buckled near end of column.
	16	15	40	12.5	9.1	:	35,490	Channels buckled near end of column.
	13	15	41	9.7	5.6	:	35,550	Filled by direct compression.
	14	15	41	9.6	8.8	:	35,550	Channels buckled near end of column.
	372	20	49	10.	8.5	:	31,698	Plates buckled between rivets.
	371	20	49	9.1	13.	:	31,302	Deflected perpendicular to pins.
	1640	30	62	12.	13.	:	33,740	Deflected diagonally.
	1641	30	62	12.3	13.	:	34,670	Deflected perpendicular to pins.
	1638	30	62	12.4	13.	:	31,110	Deflected perpendicular to pins.
	1639	30	62	12.7	13.	:	31,990	(By deflection in line of pins; column had an initial bend of ¼ inch in that direction.)
	469	20	52	13.	8.1	3	28,970	Deflected perpendicular to pins.
	740	30	63	7.6	7.	3½	34,340	Deflected perpendicular to pins.
	741	30	63	7.6	7.	:	33,530	Deflected perpendicular to pins.
	746	30	63	7.6	7.	:	33,910	Channels buckled.
	1111	30	53	7.5	8.2	:	34,810	Deflected perpendicular to pins.
	1112	30	53	9.6	5.6	:	35,240	Channels buckled.
	11	30	55	9.6	6.6	:	33,840	Channels buckled.
	12	30	56	12.	6.6	:	34,000	Channels buckled.
	9	30	56	12.3	5.2	:	34,360	Deflected perpendicular to pins.
	10	30	56	12.3	6.6	:	36,610	Deflected perpendicular to pins.
	1107	24	61	4.6	5.2	:	34,740	Deflected perpendicular to pins.
	1108	24	61	4.6	6.1	:	34,160	Deflected perpendicular to pins.
	1	25	64	4.7	3.6	:	35,880	Deflected perpendicular to pins.
	2	20	64	9.8	8.7	:	32,380	Deflected perpendicular to pins.
	752	20	64	16.4	12.5	:	30,220	Deflected perpendicular to pins.
	360	20	65			:	30,965	Deflected in plane of pins.

								Remarks
D	361	30	65	15.4	12.7	3½	31 494	Deflected perpendicular to pins.
A	757	29	63	10.2	4.75		31 380	Sheared rivets in bearing plates.
H	367	27.5	65	11.4	5.7		32 329	Deflected perpendicular to pins.
B	368	27.5	65	11.4	5.7	3	33 205	Deflected perpendicular to pins.
K	470	25	65	13.	8.1	3½	30 000	Deflected perpendicular to pins; column had an initial bend in that direction.
L	467	24	66	9.2	10.8		29 980	Deflected in line of pins.
L	1113	25	67	7.6	8.2		33 970	Deflected perpendicular to pins.
K	1114	25	67	12.1	9.1		33 610	Deflected perpendicular to pins.
E	*	25	67	12.0	9.1		32 940	Channels buckled near end.
L	358	28	67	17.8	9.3		34 240	Channels buckled near end.
L	359	28	67	17.2	9.3		32 074	Deflected perpendicular to pins.
M	3	23	69	9.7	9.1		32 251	Deflected diagonally.
C	4	23	69	9.8	13.		33 880	Deflected perpendicular to pins.
C	1636	30	70	12.	13.		33 660	Deflected diagonally.
K	1637	30	71	9.2	8.6		32 830	Deflected diagonally.
L	370	30	77	9.5	7.2		32 740	Deflected perpendicular to pins.
L	969	30	77	4.5	6.2		31 650	Deflected perpendicular to pins.
H	1231	30	78	4.6	5.2		30 720	Deflected perpendicular to pins.
H	1232	30	78	12.1	13.		33 820	Deflected perpendicular to pins.
	1634	30	78	12.2	13.		32 380	Deflected perpendicular to pins.
	1635	40	79	8.	7.		33 640	Deflected perpendicular to pins.
H	747	30	79	7.6	7.		32 440	Deflected perpendicular to pins.
H	748	30	79	7.6	7.		36 540	Deflected perpendicular to pins.
H	749	30	80	7.6	7.		34 120	Deflected perpendicular to pins.
	750	30	80	7.2	7.2		33 410	Deflected diagonally.
L	1115	30	80	12.5	8.		33 390	Deflected in plane of pins.
L	1116	30	82	12.5	4.6		34 390	Deflected perpendicular to pins.
H	58	30	83	7.75	7.6		32 160	Deflected perpendicular to pins.
L	1635	30	83	10.	7.6		32 140	Channels buckled near end.
	6	30	83	9.	5.6		31 610	Deflected perpendicular to pins.
H	468	30	82	9.2	10.8	3½	31 340	Deflected perpendicular to pins.
K	357	37	87	11.4	5.7	3	27 750	Deflected perpendicular to pins; column had initial deflection of 1 inch in that direction.
K	356	37	87	11.4	5.7		34 130	Deflected perpendicular to pins.
K	1229	35	89	4.7	6.2		31 930	Deflected perpendicular to pins.
K	1230	35	89	4.7	5.2		32 000	Deflected perpendicular to pins.
H	1653	37	93	19.5	9.4		29 186	Deflected perpendicular to pins.
H	1623	35	93	8.	3.8		29 947	Deflected perpendicular to pins.
H	1124	35	35	7.	3.8		31 370	Deflected perpendicular to pins.

TABLE No. 5.—(Continued.)

Set.	Test Number.	Ratios. $\frac{l}{d}$	Ratios. $\frac{l}{r}$	Area of Section. Square Inches.	Pin Bearing. Square Inches.	Diameter of Pin. Inches.	Ultimate Strength per Square Inch. Pounds.	Mode of Failure.
F	1645	40	96	16.2	10.	3½	29 020	Deflected in plane of pin.
F	1650	40	96	16.2	10.	"	27 910	Deflected perpendicular to pins.
A	755	30	96	10.	8.6	"	25 160	Deflected perpendicular to pins.
A	756	30	96	10.	8.6	"	21 106	(Deflected perpendicular to pins; strength reduced by bending of pin.
L	26	35	96	9.3	9.2	"	32 140	Deflected perpendicular to pins.
L	27	35	96	9.6	9.2	"	29 380	Deflected perpendicular to pins.
D	751	30	98	16.	13.	"	26 430	Deflected perpendicular to pins.
D	1642	30	98	16.3	13.	"	22 540	Deflected perpendicular to pins.
C	1651	40	98	21.	12.	"	25 770	Deflected diagonally.
C	1652	40	98	20.6	12.	"	25 990	Sheared rivets in bearing plates.
C	354	40	98	9.2	8.75	"	28 950	Triple flexure.
C	355	40	98	9.4	8.75	"	29 870	Deflected perpendicular to pins.
H	466	37.5	100	7.75	7.8	3	26 000	Deflected perpendicular to pins.
H	463	40	102	4.7	4.2	3¼	25 000	Deflected perpendicular to pins.
H	1117	40	102	4.6	5.2	"	29 180	Deflected perpendicular to pins; weakened by separators.
H	1118	40	102	7.8	5.2	"	30 990	Deflected perpendicular to pins.
J	24	40	107	7.-	9.4	"	31 350	Deflected perpendicular to pins.
J	25	40	107	7.7	7.-	"	27 850	Deflected perpendicular to pins.
J	1643	40	107	7.7	7.-	"	30 840	Deflected in plane of pins.
J	1644	40	107	7.7	7.7	"	30 770	Deflected diagonally.
J	1649	40	107	4.6	6.2	"	31 610	Deflected in plane of pins.
H	1119	45	115	4.7	6.2	"	29 870	Deflected diagonally.
K	1120	45	115	7.8	9.4	"	30 590	Deflected perpendicular to pins.
H	22	46	120	7.8	9.4	"	31 050	Deflected perpendicular to pins.
K	23	45	120	4.7	3.5	"	24 850	Deflected perpendicular to pins.
H	1121	50	128	4.6	3.6	"	26 920	Deflected perpendicular to pins.
K	1122	50	128	4.7	4.2	"	23 350	Deflected perpendicular to pins.
H	464	50	128	9.7	8.6	3	25 360	Deflected perpendicular to pins; initial bend of 0.3 inch.
A	753	40	128			3½	15 260	Deflected perpendicular to pins.
							19 380	

	No.						Strength	Remarks
A	754	40	128	9.7	8.6	3½	16 220	Deflected perpendicular to pins.
D	1046	40	131	16.2	13.	:	19 700	Deflected perpendicular to pins.
D	1647	40	131	16.1	13.	:	17 570	Deflected perpendicular to pins.
D'	20	55	141	4.7	5.2	:	21 850	Deflected perpendicular to pins.
D'	21	55	141	4.7	5.2	:	29 810	Deflected perpendicular to pins.
D'	18	60	154	4.7	7.	:	14 740	Deflected perpendicular to pins.
D'	19	60	154	4.7	7.	:	15 900	Deflected perpendicular to pins.

UNSYMMETRICAL SECTIONS.—PIN CONNECTED.

	No.						Strength	Remarks
N	1630	29	75	17.6	9.3	3¼	26 190	Deflected perpendicular to pins. Pin in center of gravity of section.
N	1631	29	76	17.2	9.3	:	29 160	Deflected perpendicular to pins. Pin in center of gravity of section.
K	352	29	77	12.6	9.1	:	27 668	Deflected perpendicular to pins. Pin in center of gravity of section.
K	353	29	77	12.8	9.1	:	30 596	Deflected perpendicular to pins. Pin in center of gravity of section.
N	1632	29	78	17.6	9.3	:	17 420	Deflected perpendicular to pins. Pin in center of channels.
N	1633	29	78	17.7	9.3	:	17 240	Deflected perpendicular to pins. Pin in center of channels.
K	350	29	79	12.5	9.	:	16 242	Deflected perpendicular to pins. Pin in center of channels.
K	351	29	79	10.8	9.	:	19 207	Deflected perpendicular to pins. Pin in center of channels.

SQUARE ENDED COLUMNS.

	No.						Strength	Remarks
E	359	29	45	17.			34 950	Plates buckled between rivets.
E	380	29	47	17.8			35 595	Triple flexure.
E	357	29	47	12.1			31 722	Plates buckled between rivets.
E	358	29	45	4.4			33 564	Plates buckled between rivets.
E	1060	25	51	4.7			36 720	Channels buckled.
F	346	21	51	15.7			35 390	Channels buckled.
F	345	21	61	13.8			32 846	Plates buckled.
F	344	21	51	21.			35 050	Plates buckled.
							33 082	Plates buckled.

* See Plate XXVII

TABLE No. 5.—(Continued.)

Set *	Test Number.	Ratios. $\frac{l}{d}$	Ratios. $\frac{l}{r}$	Area of Section. Square Inches.	Pin Bearing. Square Inches.	Diameter of Pin. Inches.	Ultimate Strength per Square Inch. Pounds.	Mode of Failure.
G	349	21	51	21.5	......	......	33 061	Triple flexure.
F	342	30	75	15.7	......	......	33 000	Deflected in plane A B.
F	344	30	75	15.6	......	......	34 505	Plates buckled.
G	341	30	75	21.2	......	......	33 019	Deflected in plane A B.
G	343	39	76	21.5	......	......	33 943	Deflected in plane A B.
N	337	30	76	17.	......	......	34 279	Deflected diagonally.
N	338	30	77	17.4	......	......	33 333	Deflected diagonally.
K	339	29	79	12.6	......	......	32 646	Deflected upwards.
K	340	25	79	12.7	......	......	33 962	Deflected upwards.

FORKED ENDS OF OPEN POSTS.—PIN CONNECTED AT ONE END.

Set *	Test Number.	$\frac{l}{d}$	$\frac{l}{r}$	Area of Section. Square Inches.	Pin Bearing. Square Inches.	Diameter of Pin. Inches.	Total Ultimate Strength.	Mode of Failure.
P	391	......	......	11.5	6.4	3½	383 000	Pin ends bent inwards.
P	392	......	......	11.5	6.4	..	415 000	Pin ends bent inwards.
R	393	......	......	11.4	4.9	..	302 600	Ends buckled.
R	394	......	......	11.8	4.9	..	302 200	Ends buckled.

* See Plate XXVII.

www.ingramcontent.com/pod-product-compliance
Lightning Source LLC
Chambersburg PA
CBHW022144090426
42742CB00010B/1385